A primary aim of *Understanding Ageing* is to dispel the view that ageing is a major unsolved problem in biology. The adult organism is maintained in a functional state by at least ten major mechanisms, which together comprise a substantial proportion of all biological processes. It can be shown that these maintenance mechanisms ultimately fail because the evolved physiological and anatomical design of higher animals is incompatible with continual survival. Studies with different mammalian species strongly indicate that cellular maintenance mechanisms are more effective in long-lived animals than in those with a short lifespan. The investment of metabolic resources into rapid growth and reproduction diminishes the resources available for maintenance, whereas this is much more effective in slowly-breeding animals. The eventual breakdown of maintenance leads to many age-related diseases, and the study of these in humans is documented by a huge biomedical literature, which is not generally regarded as being within the province of gerontology. The reality is that this information provides much insight into the changes in cells and tissues that accompany ageing. To understand in more detail the origin of all these diseases demands future research into the failure of maintenance at the cellular and molecular level. This in turn will make it easier to devise preventative measures.

This book is thought provoking because it outlines a new approach to a fundamental biological process. It will appeal to all students and researchers who are interested in ageing, whether they are working in the clinical or basic research sphere.

DEVELOPMENTAL AND CELL BIOLOGY SERIES
EDITORS
P. W. BARLOW   D. BRAY   P. B. GREEN   D. L. KIRK

UNDERSTANDING AGEING

Developmental and cell biology series

SERIES EDITORS

Dr. P. W. Barlow, *Long Ashton Research Station, University of Bristol*
Dr. D. Bray, *Department of Zoology, University of Cambridge*
Dr. P. B. Green, *Department of Biology, Stanford University*
Dr. D. L. Kirk, *Department of Biology, Washington University*

The aim of the series is to present relatively short critical accounts of areas of developmental and cell biology where sufficient information has accumulated to allow a considered distillation of the subject. The fine structure of cells, embryology, morphology, physiology, genetics, biochemistry and biophysics are subjects within the scope of the series. The books are intended to interest and instruct advanced undergraduates and graduate students and to make an important contribution to teaching cell and developmental biology. At the same time, they should be of value to biologists who, while not working directly in the area of a particular volume's subject matter, wish to keep abreast of developments relevant to their particular interests.

OTHER BOOKS IN THE SERIES

UNDERSTANDING
AGEING

ROBIN HOLLIDAY, F.R.S.

Chief Research Scientist, CSIRO Division of Biomolecular Engineering,
Sydney Laboratory, Australia

CAMBRIDGE
UNIVERSITY PRESS

Published by the Press Syndicate of the University of Cambridge
The Pitt Building, Trumpington Street, Cambridge CB2 1RP
40 West 20th Street, New York, NY 10011-4211, USA
10 Stamford Road, Oakleigh, Melbourne 3166, Australia

First published 1995
Reprinted 1996

Library of Congress Cataloging-in-Publication Data
Holliday, R. (Robin), 1932–
Understanding ageing / Robin Holliday.
    p. cm. – (Developmental and cell biology series)
Includes bibliographical references and index.
ISBN 0-521-41788-0. – ISBN 0-521-47802-2 (pbk.)
1. Aging – Physiological aspects. I. Title. II. Series.
QP86.H65 1995
574.3´72 – dc20   94–11727
                                                              CIP

A catalog record for this book is available from the British Library

ISBN 0-521-41788-0 hardback
ISBN 0-521-47802-2 paperback

Transferred to digital printing 2001

Contents

9. A better understanding of ageing 139

Preface

It is very commonly believed that the causes of ageing are unknown. For an inaugural professorial lecture, P. B. Medawar (1952) chose the topic 'An unsolved problem in biology', and this wide-ranging discussion of ageing has been influential and frequently cited. Since that time much has been learned about ageing, but most in and outside the field of gerontology would believe that the mechanism or mechanisms involved remain elusive.

My own interests in ageing began in the mid-1960s, and in the following two decades I had the same viewpoint, because it seemed that progress in research worldwide was extremely slow, so the unsolved problem remained. More recently, I have become more optimistic and have come to believe that the biological basis of ageing is now well understood. This is in part due to a consideration of the biological reasons for ageing, from an evolutionary standpoint. Also, the evolved physiological and anatomical design of mammals is well understood. The changes that occur during human ageing are documented by an enormous range of studies on age-related pathologies. A great deal is known about the many homeostatic, repair or maintenance mechanisms, which preserve the adult in a functional state for a given length of time and allow it to propagate itself. All this comprises a substantial part of biology. I believe that to understand ageing one need only look at and appreciate this vast body of existing knowledge. Those who believe ageing is a mystery cannot see the wood for the trees around them. The trees are described in considerable detail, but by concentrating on these details, many miss what the collection of trees actually comprises. I stress that the understanding to which I refer is at the broad biological level, not at the level of fine detail, which will come from future cellular, molecular and genetic studies. One cannot overemphasise the importance of such studies, because they are essential if we are ever to understand the origins of the age-related diseases. In countries with good health care where expectation of life is high, the treatment of the diseases of the aged consumes a greater and greater proportion of the total budget. Successful research in cellular and molecular aspects of ageing will inevitably improve this situation.

One reason for confusion about the causes of ageing is the continual polarisation between so-called programme theories of ageing and those that are usually described as 'wear and tear', stochastic or error theories. These have commonly been regarded as opposing viewpoints, but I no longer believe this to be true. The lifespan of the organism is determined by its genes, the products of those genes and the interaction with the environment. The genes are responsible for all those features of organisms that have been studied with success, and that provide the basis for our understanding of ageing. In this wide context, one can include development, regulation and programme, as well as wear and tear, or molecular defects. In any realistic approach, the distinction between programmed and stochastic events continually breaks down.

There is also the contentious question of single or multiple causes of ageing. Many who have proposed specific theories tend to believe their theory is the one that *explains* all of ageing. I take a different viewpoint: I believe many existing theories have some validity, and that a more global interpretation that encompasses specific theories is the most appropriate. This implies that there are multiple causes of ageing, but also multiple interactions between components of cells, or between different types of cells. In such a situation, a single molecular defect may have a wide range of effects. This interpretation is not new, and previous publications that come to similar conclusions (Olson 1987; Holliday 1988a, 1992) serve as summaries or abstracts of this book.

I am not attempting to review the field of gerontology, that is, the whole study of ageing. Many such reviews exist, most recently that of Finch (1990), which lists over 3500 references. There are, in addition, innumerable volumes that provide reviews and discussions of particular topics. Indeed, the field is extremely well endowed with review material. Instead, this book presents a particular point of view and argument. It draws on information that is relevant, and ignores a great many studies of ageing. Many of these studies are interesting and important; they may, for example, document the change of a particular physiological or biochemical parameter with age. Nevertheless many such observations are at present not interpretable. In my opinion, they comprise part of the level of fine detail, and their significance will be understood only in the future. In developing the major theme, I draw only on material that I believe is relevant to *Understanding Ageing*.

Author's note

As I indicated in the Preface, this book is not a review of ageing. It will also become apparent to the reader that the level of documentation and citation of previous publications is quite uneven. Sections of the book that deal with general or biomedical topics, such as the pathological changes that often accompany human ageing, are almost absent of references. Much of this information is available in standard medical texts or monographs. In other parts of the book I cite wherever possible previous reviews or collections of publications that cover a particular topic. There are innumerable reviews of different aspects of ageing, including three editions of the *Handbook of the Biology of Aging* (Van Nostrand Reinhold, 1977, 1985; Academic Press, 1990) and the CRC series on ageing, which comprises at least 12 volumes on diverse topics in gerontology. The sections that are most thoroughly documented with references are those that have over the years been the focus of my own research and interests: those related to cellular ageing, the accuracy of information transfer between macromolecules and the evolution of ageing. Some of this work has been previously reviewed (Holliday 1984c, 1986b). I am well aware of the more limited discussion of equally important topics, such as the accumulation of abnormal proteins in various tissues, or neuroendocrine interactions. To keep the text as readable as possible, I have assigned specific technical or theoretical points to a series of appended Notes, supported with further references and, in some cases, with illustrations. Finally, I wish to apologise in advance to any authors whose work I may discuss without an appropriate citation, and also for any errors in citations.

Acknowledgements

This book was written after thirty years of experimentation and discussions about ageing, and I wish to thank all the various colleagues who contributed to this research. The work began with *Drosophila*, continued with fungi, and also included experiments with mice and *Escherichia coli*. The main body of work was devoted to the study of the *in vitro* ageing of human cells. Those who collaborated with me in all these experiments, listed in the approximate chronological order in which they joined in the research, are Brian Harrison, Cynthia Lewis, Gill Tarrant, Katherine Thompson, Ian Buchanan, Tony Stevens, Steve Fulder, Lily Huschtscha, Stuart Linn, Mike Kairis, Vince Murray, Clive Bunn, Keen Rafferty, Robert Rosenberger, Khash Khazaie, Alec Morley, Simon Cox, Robert Stellwagen, Gillian Foskett, John Menninger, Suresh Rattan, Toshiharu Matsumura, Jacqueline Hunter, Farooq Malik, Gayle McFarland and Alan Hipkiss. I have also greatly appreciated collaboration with Tom Kirkwood in several theoretical studies, and many discussions with Leslie Orgel. Others who have contributed in one way or another include Arnold Burgen, Zhores Medvedev, Rita Medvedeva, Geoff Banks, Ad Spanos, Steve Sedgwick and Jonathan Gallant.

I thank Jonathan Slack, the series editor, who invited me to contribute a book on ageing. The final text was greatly improved by the very helpful comments of Leslie Orgel, John Grimley Evans and Geoffrey Grigg. I also thank Lily Huschtscha and Gayle McFarland for reading the text and pointing out errors. I am particularly grateful for all the hard work that was expertly carried out by Anne McGill, who prepared all the material for publication. I also thank Caroline Holliday who typed some drafts of chapters; Ann Neville and Suzanne Anderson for various library searches; Jane Michie for drawing figures and Louise Lockley for preparing prints. Unpublished scientific data were kindly provided by Marvin L. Jones (San Diego Zoo), Seiji Ohsumi (Institute of Cetacean Research), Kevin Johnston (Melbourne Zoo), Virginia Landau (Jane Goodall Institute), and Pamela Pennycuik.

Finally, I thank the Medical Research Council, U.K., which supported all the work on ageing at the National Institute for Medical Research, Mill Hill, and the Commonwealth Scientific and Industrial Research Organisation (CSIRO) of Australia, which has supported continuing research at their Sydney Laboratory, Division of Biomolecular Engineering, North Ryde, and also made it possible for me to write this book.

1

Introduction

The growth and reproduction of all organisms is dependent on a source of energy and other essential requirements from the environment. These resources for growth are never unlimited, from the simplest life forms to the most complex. Under good conditions, bacteria can divide every 30 minutes or so. From one cell 24 hours' growth produces 2^{48} or 10^{16} cells, with a total mass of about 30 kilograms. It is obvious that in any natural environment such exponential growth cannot be sustained. In such an environment, cell division will continue until nutrients become limiting and cells enter what is generally referred to as 'stationary phase'. There are different possible fates for such cells. They may die from prolonged starvation or dehydration, they may provide food for other organisms (such as nematodes, which feed on soil bacteria) or they may renew growth if a supply of energy and nutrients becomes available. Bacteria illustrate the general demographic principle that, for many environments, the number of a specific type of organism approximates to a steady state. In a given volume of soil, for example, the number of a particular species of bacteria may fluctuate between larger and smaller populations, but one can make the broad generalisation that numbers will remain roughly constant over quite long periods of time.

Darwin based his principle of natural selection on the fact that the reproductive potential of organisms can never be realised, but those that are better adapted to a particular environment use the resources more successfully and have increased reproductive fitness. In a bacterial population a cell resistant to an antibiotic may replace one that is sensitive; nevertheless even the better adapted population is still constrained by the limits to reproduction imposed by the environment. The same principle applies to more complex organisms, whether they multiply asexually or sexually. Plants frequently have means of vegetative propagation and are in this respect potentially immortal. Although a single vegetative clone by its capacity for reproduction could in theory take over or dominate an environment, in practice there are limits to such proliferation, and single plants produced vegetatively have a lifespan imposed on them. These individual plants may die from drought, from infestation by pathogens, or they may be eaten by animals. Some simple animals, such as hydra, can also reproduce asexually by budding off new individuals, and the evidence suggests this can continue indefinitely. Other coelenterates, such as sea anemones, have the ability to regenerate damaged or lost parts, and have very long recorded lifespans in aquaria.

The combination of sexual and asexual reproduction, seen in many plants and some simple animals, has been lost in the higher animals. It is in these organisms that Weismann's distinction between an immortal germ line and a mortal soma becomes clear-cut. Nevertheless in the evolution of the invertebrates there are interesting transitional states. Flatworms have the three cell lineages, ectoderm, endoderm and mesoderm, characteristic of all higher animals, but they lack a body cavity surrounding the gut. The survival of the free-living flatworms (planarians) has been studied experimentally (Child 1915; Sonneborn 1930). It has been shown that bisection of an animal into head and tail regions is followed by regeneration of the tail to form a complete organism, and this process can be repeated as long as the investigator has patience. The head, however, does not have the same power of regeneration, and it does not survive for very long, even if provided with an optimal environment. It seems very probable that the tail of the animal has a pool of potentially immortal cells that are totipotent and can regenerate all parts of the animal: cells that are fully differentiated would have limited survival time, but in a normal animal can be continually replaced from the totipotent pool. This interpretation is strongly supported by experiments with X-rays. An appropriate dose of radiation kills the pool of totipotent cells, but leaves the rest of the animal essentially undamaged. Such an animal can feed and survive for 60 or 70 days, but it then shows signs of senescence and subsequently dies (Lange 1968). Experiments with flatworms are instructive because they illustrate the mortality of specialised cells, and the immortality of a minority of other cells.

Animals more complex and advanced than flatworms have very variable powers of regeneration. There are species in many taxonomic groups in which the adult is of defined size and the somatic cells are incapable of further division. Examples are nematodes and many insects, which also have a well-defined maximum lifespan. In many other species, growth continues throughout adult life, and in general ageing is much less well defined. Some reptiles and fish continually increase in size and can achieve very long lifespans, such as the whale shark, crocodile or giant tortoise. There are vertebrates that retain considerable powers of regeneration; for example, some amphibia and reptiles can regenerate severed nerves, limbs or tail.

The most highly-evolved vertebrates are the warm-blooded mammals and birds, and in these we almost always find that adults have a well-defined size and lifespan. These animals have tissues consisting of non-dividing cells, such as those in the brain or heart, but there are also stem line cells, which continually replace those with limited survival time, such as cells in the skin, blood, or the lining of the gut. Although wound repair is often effective in mammals and birds, powers of regeneration are limited.

All these species, whatever their powers of regeneration, have a soma with finite survival time. The strategy for the survival is development to adulthood, investment of resources in reproduction, but insufficient resources for continual maintenance of the soma. The evolution of this strategy is the consequence of the basic demographic principle that has been outlined. Most natural environments are hazardous with limited availability of food, and under these circumstances mortality is high. Maximum Darwinian fitness is achieved by the successful production of surviving offspring rather than by investing resources in the long-term maintenance of the adult. Darwin's theory of natural selection was in part derived from Malthus's realisation that the full reproductive potential of any species cannot be achieved under normal conditions, because if that potential were fulfilled, population size would simply increase exponentially. The hostility of the environment provides not only the selective force for better adapted individuals, but also explains the irrelevance of the long-term survival of the soma. In terms of population genetics, one can state that natural populations are age-structured, with a continuously declining proportion of older individuals (Charlesworth 1980; Rose 1991). Most offspring are produced by the youngest adults, and on a population basis, fewer and fewer offspring come from adults of increasing age. The widespread existence of ageing in animals can be understood in this evolutionary framework. The relationship between reproductive potential of an organism and its maximum lifespan is discussed in more detail in Chapter 7.

So far, ageing has been referred to as a limit to survival of the soma or body. There are two fundamental parameters for ageing: first, the maximum lifespan, and second, the expectation of life, or average lifespan. These parameters can only be studied in populations of animals, not in an individual animal. As we have seen, in a natural environment the death of most animals is due to starvation, drought, predators or disease, and the expectation of life is therefore largely dependent on the environment. The maximum lifespan is seen only in protected environments, which are broadly constant over time. Examples of these are laboratory cages for experimental animals, zoos and those societies that provide adequate food and health care for people, or for their domesticated animals. A cohort or population of animals in such environments produces a survival curve such as Figure 1.1D. There is usually some infant mortality, followed by a long period of survival with sporadic deaths. Thereafter the 'force of mortality' increases, and corresponds to the increasing slope to the survival curve. Finally, a few individuals are left who are approaching or reach the maximum lifespan. Obviously the shape of the curve is influenced by environmental factors, and also by the natural genetic variability in any population. It is important to note, however, that inbred mice, which have the same or extremely similar genotypes, kept under uniform conditions in laboratory cages, have survival curves very similar to that shown by Figure 1.1D, and examples are shown in Figure 1.2. This result shows that there are intrinsic events that strongly influence lifespan that are

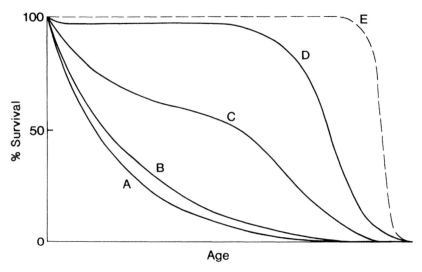

Figure 1.1. Generalised survival curves. A, exponential (constant mortality with time); B, the survival of small animals, such as mice, in a natural environment; C, the survival of large, slowly-breeding animals, such as elephants, in a natural environment; D, the survival of experimental animals in a laboratory environment, or of humans in societies with good health care; E, hypothetical survival of a genetically uniform population in a constant environment – no such survival curve has ever been recorded.

neither environmental nor genetic. Such events are probably random or stochastic, and this possibility is discussed in more detail in Chapter 4. The shape of the survival curve provides strong evidence against a genetically-determined clock or strict programme for ageing, which for inbred animals would produce the survival curve indicated by the dashed line Figure 1.1E. In many species that have been experimentally examined (especially genetically uniform mammals, insects and nematodes) such a survival curve has never been recorded.

Senescence becomes most obvious when the force of mortality increases, and its major features are described and discussed in later chapters. The terms 'ageing' and 'senescence' are usually used more or less interchangeably, and this practice is followed here. Both are characterised by progressive or accelerated changes in the tissues and organs of the body, leading inexorably to death. Superficially it appears that these changes are induced or are in some way switched on after a long period of normal adult life. In reality, the shape of the survival curve is the same as that produced by 'multiple-hit' kinetics. If there is a constant probability of an event occurring, but only several accumulated events produce a phenotypic change, then the probability of that phenotype occurring is initially very low but becomes much higher later in life. This is the same argument used by epidemiologists to explain late-onset carci-

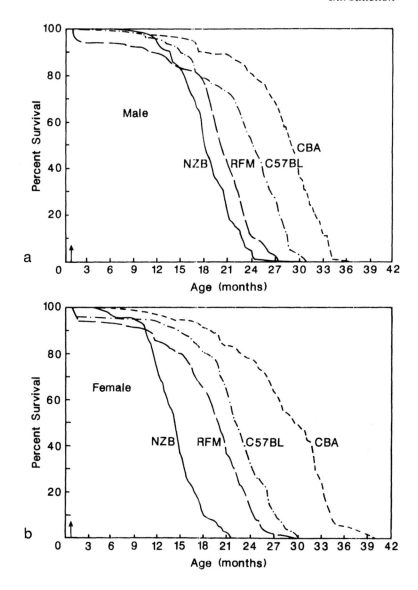

Figure 1.2. Survival curves of various inbred strains of mice. (Redrawn from Zurcher et al. 1982.)

nomas, since several sequential events, such as mutations, will produce the age-related increase that is seen (see Chapter 8).

The force of mortality can be translated into a quantitative parameter, and one that has been often discussed is the Gompertz function (see Note 1.1). In

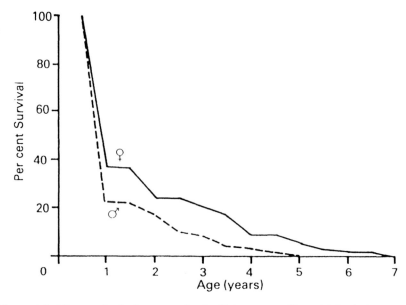

Figure 1.3. The survival of grey squirrels (*Sciurus carolinensis*) in their natural environment. (Reproduced with permission from Gurnell 1987.)

more general terms, the quantitative aspects can be illustrated by dividing the lifespan into equal units (days, weeks or years depending on the organism), and then considering the probability of dying within each successive unit. Thus the force of mortality is usually quite high in infancy, is at a low level in adulthood and then rapidly increases as the shape of the survival curve becomes steeper. This change in the force of mortality associated with ageing is in contrast to the survival curve seen when the probability of dying is constant with time. This can be most easily understood in the context of inanimate objects, such as the survival of coffee cups in a café, or glass pipettes in a laboratory. Their survival is exponential, producing the curve A in Figure 1.1. The rate of decay of radioactive isotopes is also constant with time and follows the same kinetics. The lifespan of other inanimate objects may follow 'multiple-hit' kinetics. Imagine, for example, plastic graduated pipettes which do not break when dropped and are continually reused. Their useful lifespan depends on the number of scratches accumulated, or perhaps their increasing opacity.

In natural environments, the survival curve of populations of animals lies somewhere between the curves A and D in Figure 1.1. An example is shown in Figure 1.3, which documents the survival of the grey squirrel in its natural environment. The relatively short survival time is striking, in view of the fact that this animal has survived in a zoo environment for 21 years (Jones 1982). Small, rapidly-breeding ground animals, such as mice, have survival curves scarcely different from exponential (Fig. 1.1B), whereas large, slow breeding animals such as elephants or large whales must necessarily have longer average lifespans to maintain their numbers, and their survival might approximate to curve C. The relationship between rate of development, reproductive potential and lifespan is discussed in more detail in Chapter 7.

The major features of senescence and ageing obviously vary between major taxonomic groups. Not all these features of ageing are discussed in subsequent chapters; instead, attention is focused on the ageing of mammals. During the ageing of mammals, and of many other taxonomic groups, deterioration of structure and function is seen in a wide variety of organs and tissues as the force of mortality increases. It is remarkable that very similar changes are seen in different mammalian species irrespective of their maximum lifespan, and it is therefore important to undertake comparative studies using species with different lifespans (see Chapter 7). Some of the best known features of ageing include loss of elasticity of the skin, decline in muscular strength, loss of hair pigment, disorders in the endocrine system and decline in the immune response. The circulatory system deteriorates with loss of elasticity of arteries and the accumulation of atherosclerotic plaques on the inner wall. The normal functions of internal organs, such as the kidney and liver, gradually decline. The brain loses neurones, and histological abnormalities accumulate, such as lipofuscin (the so-called age pigment), tangles or plaques. The retina of the eye often becomes damaged, and the lens may form cataracts. More generally, there is a decrease in the organism's ability to respond to stress imposed by the environment.

Pathologists regard the diseases that become more frequent during ageing in much the same way as all diseases, and they frequently make a distinction between so-called natural ageing and age-related disease. The main reason for this is the uneven distribution of age-related diseases. Alzheimer's disease is regarded as a pathological condition because many old people retain quite normal mental abilities. Similarly, cardiovascular disease, diabetes, cancer and so on are regarded as distinct pathologies that are not part of ageing per se. This view is mistaken, because to understand age-related changes it is necessary to consider ageing populations, not the ageing of one individual. This is illustrated in Figure 1.4, which demonstrates the distribution of an age-related disease in a population at the time of diagnosis. It also indicates the average expectation of life of the individuals in the population. The distribution of lifespans (A) is to the left of the median of the distribution frequency of the disease (B). This means that only a *proportion* of elderly people are diagnosed as having the disease. The longer the individual lives, the greater is the probability of having the disease; but this never rises to 100%, as not all individuals reach a sufficiently-advanced age. For each different disease the distribution will be different, and in some cases the curve may be skewed (not shown). The distribution of the disease in question will be broad in some cases, such as cardiovascular diseases in humans, and narrower in others, such as senile

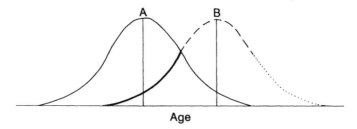

Figure 1.4. Curve A is a generalised distribution curve of the lifespans of individuals in a society with good health care. B is a hypothetical distribution curve of the age of diagnosis of a particular age-related disease. The part of the curve shown dotted can never be measured, because individuals do not survive long enough. The distributions show (1) that many individuals die without having the disease, (2) that the probability of contracting the disease increases with age, and (3) that the incidence of the diagnosed disease increases more than linearly with time, and may approximate to an exponential increase. The distribution B is different for each age-related disease, and could be skewed rather than symmetrical.

cataracts. The curves also illustrate another important feature of ageing populations: the rate at which specific deleterious changes are seen does not increase linearly with time; instead, the rate of increase often accelerates. This is also shown in Figure 1.4 where, as the individual approaches advanced age, the probability of acquiring a particular disease increases more than linearly.

The assumption is therefore made in this book that the whole spectrum of changes seen in populations of elderly individuals are all part of ageing per se. This means that the vast knowledge that has been accumulated from the study of age-related pathological conditions in humans is a major component of gerontology. This theme is taken up again in Chapter 8, where common age-related pathological conditions are briefly reviewed. Another major theme in this book is that the deterioration and disease of old age are a direct consequence of the evolutionary design of higher organisms. Ageing is of very ancient origin, and the anatomical and physiological features of most animals are simply incompatible with indefinite survival. The evolved design of mammals in relation to the inevitability of ageing is discussed in Chapter 2.

Although ageing may be inevitable, longevity can be modulated or controlled. Complex organisms survive as long as they do through a wide range of cell and tissue maintenance mechanisms. These preserve the adult animal in a vigorous normal physiological state, and allow reproduction over a considerable period of the total lifespan. The range and sophistication of the various maintenance mechanisms are reviewed in Chapter 3. It is argued throughout this book that *ageing is due to the eventual breakdown of maintenance, and that this breakdown is the inevitable consequence of the evolved anatomical and physiological design of the organism.*

Many theories of ageing have been proposed in this century, and most suggest a single major cause of ageing, rather than multiple causes. It is significant that each of these theories discusses, in effect, the failure of a specific maintenance system or mechanism. For example, the 'immunologic' theory of ageing proposed by Walford (1969) is based on the deterioration of the immune system, which is a fundamental maintenance mechanism of the body. Theories and mechanisms of ageing and their relationship to maintenance are discussed in Chapter 4.

It is a truism that ageing is genetically determined, because each species has a characteristic well-defined lifespan. It is often assumed that this genetic determination means there must be a strict programme for ageing: thus mice are programmed to die after about three years and humans after 70–100 years. This causes confusion because the word 'programme' implies a specific mechanism, such as a clock, whereas the reality may be far more complicated. Every protein in the body is genetically determined, because the sequence of bases in the DNA of a structural gene specifies the amino acid sequence of the protein. The protein may be an enzyme, a component of a membrane or organelle, or have either an intracellular or an extracellular structural role. The properties of cells and tissues are largely determined by proteins, and the mechanisms that maintain the integrity of the body throughout adulthood all depend on proteins. Proteins can be stable or unstable, depending on their own amino acid sequences, on their local physiological environment, and on the properties of proteases that may recognise and degrade them. So it is a fact that the totality of cellular metabolism, in all its aspects, is genetically determined. Therefore, the genetic determination of ageing can operate through direct effects on this metabolism.

This is the key to the remarkable feature of ageing previously mentioned, namely, that the major changes seen are very similar in short-lived and long-lived mammals. For example, the cross-linking of collagen is a well-known 'biomarker' of ageing, and it occurs very much faster in short-lived animals than long-lived ones, although the underlying chemistry is the same. Similarly, the crystallin of the eye lens, which is laid down early in development and never replaced, is much more stable in long-lived animals than short-lived ones. In both cases old individuals can develop cataracts, but the time-span may differ by 30-fold.

How does a programme for ageing simultaneously affect collagen, crystallin, neurones, arteries and innumerable other components of the body? Clearly the question becomes meaningless, because the programme must depend on the structure and properties of the individual components of the body, and these are ultimately determined by the genome, which evolved over an enormous time-span. Nevertheless, there are specific adaptations in some species that terminate life at a particular time, as is reviewed in Chapter 6. These adaptations are related to the life style or the environment of the organism in question, and it is appropriate to refer to them as examples of programmed ageing. In the absence of such adaptations, the general concepts of programmed and stochastic ageing are so closely interrelated that the controversy between apparently opposing views is to a large extent artificial.

The ageing of the organism is manifested in tissues and organs, but these are made up of cells. So to understand ageing we need to consider the ageing of cells, which in turn is brought about by changes in molecules. Cellular ageing can occur in non-dividing post-mitotic cells, and also in dividing cells. In the latter it is necessary to distinguish between the ageing of an individual cell and a population of cells. Orgel (1963) has drawn attention to certain fundamental features of the ageing of dividing cells. A population of bacteria or yeast cells will continue to divide indefinitely if provided with a sustaining environment. Clearly these populations are immortal, but the situation is different if we consider individual cell lineages, as the following 'thought experiment' shows. When a bacterium divides, one cell is selected at random and allowed to divide. Again one cell is selected at random and allowed to divide, and so on. With this random selection of cells, the lineage will inevitably die out, because sooner or later a lethal event (such as a mutation) will occur. Scrutiny of real populations of bacteria will reveal occasional dead cells, but of course the population keeps growing. Thus we conclude that individual cell lineages are finite within populations that have infinite growth potential. The population as a whole will continue to grow provided the probability of any daughter cells surviving is greater that 0.5. To illustrate this, Orgel (1973) considered another experiment, using a chemostat. Such populations are in a steady state, with nutrient being supplied and cells and spent medium being withdrawn at constant rates. If the chemostat is now placed close to a radiation source, bacteria will be killed by induced lethal events. When the frequency of killing becomes greater than 50% for each new daughter cell, then the population will decline and die out. (Ignoring, for the purpose of the argument, the selection of occasional radiation-resistant mutants.) At any lower radiation dose, the population will continue to grow, albeit at a lower rate than the unirradiated one.

The same applies to many mammalian cell lines, such as the HeLa cell line, which can be grown continuously as a population. Such cell lines have a significant proportion of non-dividing cells, usually in the range 10–30%, but these are continually selected out, and the population keeps growing. It was originally thought that any population of dividing avian or mammalian cells could be grown indefinitely in culture, but this was later shown to be untrue (see Note 1.2). Human diploid fibroblasts, which are normal cells derived from tissue biopsies, grow as a population at a constant rate but eventually slow down and cease growth after 50–70 population doublings. This was first clearly documented in Leonard Hayflick's laboratory (Hayflick & Moorhead 1961), and the finite growth of these cells is therefore often referred to as the 'Hayflick limit'. Hayflick (1965) suggested that the behaviour of cultured diploid cells provided an experimental model for the study of ageing at the cellular level. There are several compelling reasons for believing this is correct. In particular, comparative studies of cells from different species show that the number of population doublings achieved is directly correlated with the maximum lifespan of a donor species. These results and the general features of cellular ageing *in vitro* are reviewed in Chapter 5.

Experiments on the transfer of cells or tissues from animal to animal indicate that their lifespan *in vivo* is also finite (see Chapter 5). This is not universally accepted, and it has been claimed that stem line cells divide every one or two days throughout the lifespan of an individual, which is a far greater number of cell divisions than the Hayflick limit. However, this cannot be established with experimental certainty, because stem cells might be replaced from time to time from a quiescent non-dividing pool, which means that the number of cell divisions in any one lineage could be within the Hayflick limit (see Chapter 5).

Although dividing cells may well be important in ageing, it is more likely that post-mitotic non-dividing cells are of greater significance. Cells that are metabolically inert, such as those in spores, seeds, other dormant structures, or those stored in liquid nitrogen, survive extremely long periods of time. In contrast, cells that are active in metabolism are subject to many deleterious events; for example, a spontaneous dominant lethal mutation will end the life of such a cell. The point is similar to that made in Orgel's 'thought experiment' using non-selected bacterial lineages. Sooner or later a cell in a continuous lineage, or a cell that will never divide again, will suffer one or more deleterious events that will result in death. Since neurones never divide and are progressively lost during ageing, we know that the viability of these cells in short-lived animals, such as small rodents, is preserved or maintained much less efficiently than the viability of exactly the same type of cell in a long-lived species. It is clear that the maintenance of the organism as a whole depends in large part on the maintenance of individual cells, whether dividing or non-dividing. The challenge for the future is to understand these differences in maintenance at the molecular level.

2

The evolved anatomical and physiological design
of mammals

The recapitulation theory proposed that the development of an organism mimics the stages of evolution that lead to the adult species. In other words, ontogeny recapitulates phylogeny. The theory is no longer seriously considered, but it does illuminate a fundamental feature of complex organisms. Whereas machines are put together piece by piece, organisms start from a single cell, the fertilised egg, and through intrinsic processes produce a series of developmental stages of increasing complexity until the full-sized adult animal is formed. Nevertheless an animal has much in common with a machine; for example, the intake of energy and its conversion to output follow the same thermodynamic laws. The mechanistic aspects of movement and locomotion may be very similar. The articulation of joints and bearings have much in common, and the wings of bats and birds use the same aerodynamic principle of lift as do those of aeroplanes and gliders.

The comparison of animals and machines, like all analogies, can never be exact, but their similarities and differences are relevant to any discussion of ageing. Machines are designed and put together to produce the functional unit. The machine has no means of maintaining itself or repairing defects, except perhaps in special cases. The designer, of course, takes all steps to prevent parts wearing out, for example, by providing initial lubrication, frictionless joints, and so on, but any machine will eventually wear out unless it receives maintenance or repair from outside. This pinpoints two fundamental features of most machines. First, an unattended machine has a limited lifespan. Second, continued maintenance and repair, including replacement of parts, can provide unlimited lifespan. Vintage cars are lovingly maintained by their owners, but at very high cost. Parts are replaced as required, and, in principle, the car could still exist in its original form even if none of its original parts are retained. As everyone knows, such expert maintenance is extremely expensive, particularly as replaced parts may require special engineering. For most cars, this is too expensive; hence the popular conclusion that cars have 'built-in obsolescence'. However, the phrase is misleading in the sense that the car is not old or discarded because it is an obsolete model, but because the wear and tear it receives makes it unreliable and not worth repairing.

Complex organisms also have many structures that need constant maintenance. Unlike the machine, the organism *repairs itself.* The many repair and maintenance mechanisms are reviewed in Chapter 3. In this chapter, and also Chapter 8, features of the organism that cannot be repaired, maintained or replaced are discussed. The aim is to show that the evolved design of mammalian species necessarily results in a limited lifespan. It is tempting to apply the expression 'built-in obsolescence' to organisms that cannot survive indefinitely. However, as with cars, the word 'obsolete' is inappropriate; an organism that lived a very long time would not be obsolete. Nevertheless the phrase 'built-in obsolescence' can be useful, provided it is taken to mean limited survival time. With regard to survival of humans and domesticated animals, we have to consider the special cases of repair by medical and veterinary treatments. The relevance of this to ageing, or the prevention of ageing, is mentioned later in this chapter, but discussed more specifically in Chapter 9.

An interesting comparison of spacecraft with ageing organisms has been attributed to Alex Comfort. A spacecraft is designed to reach a particular destination, so it must be able to function properly for the necessary length of time. The craft may be specifically programmed to last the required time, for instance, by the provision of batteries that will provide power for the whole journey. The batteries may have some surplus power, but if the destination was missed, they would run out reasonably soon afterwards. The organism is also in a sense programmed to last a given length of time, and each essential component must function for at least as long. Comfort's comparison can also be considered in another way. Suppose the spacecraft has its own power from solar batteries and all its components are made, designed and assembled with the greatest possible efficiency. To minimise failure, the spacecraft is designed to last as long as possible. If this craft misses its destination, it will continue to function, perhaps for a long time, but never indefinitely. Sooner or later components will function less efficiently or fail altogether. In this case the spacecraft's journey is not programmed, but its lifetime depends on the extent of wear and tear in various components. As we shall see, organisms also age through the wearing out of essential components, which is often described as a stochastic process of ageing. In the previous chapter, it was suggested that the distinction between programmed and stochastic ageing is in large part artificial (except in special instances, which are discussed in Chapter 6), and this theme is continued in this and later chapters.

With regard to wear and tear in relation to ageing, teeth are particularly instructive. Although teeth are complex structures including a blood supply and nerves, the hard outer dentine and enamel provide the essential mechanical components used for biting, chewing or cropping. There is no mechanism for the replacement of these components. With long continued use, teeth will wear down to a point where they are unable to fulfil their function and, of course, they are also subject to decay or other types of damage. It is well known that the extent of wear of the incisor teeth of a herbivore, especially the horse, provides an estimate of the age of the animal. When a grazing animal becomes old, the teeth may be completely worn down, and in a natural environment it could then starve. This raises interesting questions, in particular, is the death of an animal from worn-down teeth a case of natural ageing? If this is so, then clearly we have a case of ageing through wear and tear. However, the lifespan of the teeth is determined by their length in the young adult. Clearly tooth length is genetically determined or programmed. Thus, we have a genetic programme specifying the length of teeth, and also constant wear and tear, which gives the teeth finite lifespan. The distinction between programmed and stochastic ageing clearly breaks down in this case. The only conclusion one can draw is that the evolved design of the teeth is sufficient to last about as long as the animal's maximum lifespan. Teeth also provide an interesting example of the similarity of part of an animal to part of a machine. Teeth cannot be repaired or maintained by the organism, but they can be very effectively repaired by the expertise of a dentist, or they can be replaced with false teeth. Similarly, components of machines that are subject to wear need to be repaired or replaced.

Some herbivores have got round the problem of tooth wear by evolving essentially a steady state mechanism. The incisors of the rabbit grow continuously at the base, so the teeth never wear out. In other animals, such as the shark family, old teeth are continually replaced by new ones. Steady state systems are also seen in other surface structures. Nails provide a good example: the continual damage to the end of the nail is made good by growth from the base, although the rate of nail growth declines during ageing (see Williams, Short & Bowden 1990). The loss and replacement of skin and hair is more complex, and varies greatly between species. In principle, the wearing of skin unprotected by hair is made good by the continual production of new cells in the epidermis. In fact, skin is subject to ageing, related in large part to changes in structural molecules such as collagen (which is discussed later in this chapter).

In the early evolution of ageing in animals, a crucial step was for formation of post-mitotic cells, which are not replaced. In Chapter 1 the regenerative capacity of flatworms (planarians) was discussed: these are animals in which differentiated cells can be replaced from a totipotent pool of cells. Another type of worm, the roundworm or nematode, is more advanced in that it has a body cavity, or coelum, which flatworms lack. Its cellular organisation is also totally different. In the development of a nematode such as *Caenorhabditis elegans,* there are strictly defined cell lineages that give rise to a constant number of cells in the adult (Sulston et al. 1983). These cells are highly differentiated and post-mitotic, and they cannot be replaced from any stem line or pool of undifferentiated cells. As a result, the nematode has a very well-defined lifespan. Its lifespan is due to the finite survival time of individual cells, although the exact relationship between cellular abnormalities and death of the whole organism is not known. Many other invertebrates also have an adult soma that consists of post-mitotic cells; insects are one of the best-known examples. Adult *Drosophila* has dividing cells in its germ line, but almost all its other cells are post-mitotic. Like nematodes, flies have a very well-defined lifespan, so not surprisingly both have been extensively used in experimental studies on ageing. It is unrealistic to expect non-dividing, metabolically active cells to survive indefinitely. There are many reasons why cells might die, and these are discussed elsewhere in this book. It is sufficient to stress here that macromolecular systems are not in equilibrium or a steady state. For example, although DNA is constantly repaired, sooner or later irreparable lethal damage would be expected. The loss of both copies of an indispensable gene in a diploid cell means that the cell dies. Also, in many cellular contexts there is a balance between the production of abnormal proteins and their removal. However, non-dividing cells active in metabolism eventually accumulate insoluble protein material, or the 'age pigment' lipofuscin, which cannot be removed by hydrolytic enzymes.

In the evolution of higher vertebrates, especially mammals and birds, very important parts of the body are also largely made up of post-mitotic cells. Neurones in the brain provide one of the best examples: they are not replaced, and the brain as a whole has limited powers of repair. A well-known textbook of pathology makes the point as follows:

> Localization of function makes the brain inherently vulnerable to focal lesions that in other organs might go unnoticed or produce only trivial symptoms. . . . This vulnerability of the brain to small lesions is compounded by its very limited capacity to reconstitute damaged tissue. There has clearly been a serious error by the celestial committee in its design of an organ that is vital for biologic survival and yet is both the most vulnerable to, and the least tolerant of, focal damage. (Morris 1989, p. 1386)

The human brain is the most complex biological structure produced by evolution on this planet, but the evolved structure is only capable of function for a lifetime. It is well known that the number of brain neurones declines with age (see Johnson 1986). Some of the pathological changes that occur in an aged brain are briefly reviewed in Chapter 8.

The eye is another structure that very clearly has limited survival time. The lens of the eye has an epithelial layer that produces early in life cells that differentiate to form the crystallin proteins to make up the lens. The lens formed early in life is largely an inanimate structure and it does not have the means to repair itself. The young lens is obviously highly transparent, and it is also elastic so that the attached muscles can change its focal length, or accommodate the lens to focus at objects at different distances. Unfortunately, both the transparency and the accommodation are not permanent. As the eye gets older the elasticity of the crystallin components declines and the ability to focus over a wide range is lost. In fact, the loss of accommodation in humans is essentially continuous from childhood. Later on the transparency of the lens declines, and this can lead to loss of vision by cataract formation. The formation of cataracts is due to post-synthetic changes in the crystallin molecules, leading to a loss of transparency. Various post-synthetic changes in proteins are a very important component of ageing in cases where proteins are long lived and either cannot be replaced or are replaced very slowly. The chemistry of some of these changes is reviewed in Chapter 4.

As well as the lens, the retina is also subject to age-related changes, which are described in Chapter 8. The major reason for its ageing is that the cells cannot be replaced, and there are good reasons why they cannot survive indefinitely. This is illustrated by the turnover of photoreceptor elements in the rods of the retina. These receptors are layered in the rod like a pile of coins, and new receptors are added at the top of the pile, nearest the light source, and removed at the base. In the human retina about 90 new receptor elements are produced per day, or roughly three million in a lifetime. The same number, having performed their function, are removed from the base. They are partly degraded, but insoluble debris in secondary lysosomes are phagocytosed by the underlying pigmented epithelial layer. Each epithelial cell is in contact with about 25 rod cells, and it has been calculated that each epithelial cell phagocytoses 2000–4000 discs per day, depending on its position in the retina (Hogan 1972). This is not a steady state situation, and with time non-degradable components accumulate, including the so-called age pigment lipofuscin, which impairs the normal function of the retinal epithelium and of the eye itself. The turnover of the receptors is not enough; to preserve the eye indefinitely would depend on the replacement of defective rod and cone cells, and other components of the retina (see also Chapter 8). This replacement mechanism never evolved.

The ear is another complex sense organ, which is terminally differentiated to provide long, but not endless, function. The cochlea contains sound-sensitive hair cells, which initially respond to a very wide range of sound frequencies. It is well known that children can hear higher frequencies than adults; indeed, the loss of ability to hear these frequencies is a good 'biomarker' for ageing. With time, these cells lose function as do those responding to lower frequencies, and there is no mechanism for their replacement. Deafness is one of the commonest conditions of old age.

Apart from those in the mammalian brain, and its sensory extensions the eye and ear, there are post-mitotic cells in other parts of the body that are also subject to damage and are not replaceable. The heart consists in large part of multinucleate myotubes incapable of division. Whereas other muscles are capable of repair through the activity of myoblasts, this is not the case for heart muscle. The continual use of the heart in the absence of ongoing maintenance inevitably leads to eventual loss of function. The valves of the heart operate approximately 2.5×10^{11} times during a human lifespan and are particularly subject to wear and tear. It is well known that they often become irreversibly calcified, and life can only be sustained by surgical replacement with artificial or pig heart valves.

Arteries are also non-renewable structures subject to irreversible damage (see Chapter 8). The wall of the artery contains long-lived proteins such as collagen and elastin, and with time such molecules become cross-linked. This leads to a loss of elasticity and hardening of the artery wall. At the same time atherosclerotic plaques accumulate on the inner wall of the larger arteries, and these can prevent normal blood circulation. Although some repair is possible in major vessels, this in turn leads to scar tissue, which is itself imperfect. The basic problem is that the lumen of an artery is simply in the wrong location for effective repair.

The problem with both the brain and the vascular system is that continuous function is necessary. When a working machine is to be repaired, it is first stopped, and restarted after the repairs are completed. This is not possible for vital organs in continual use. It *is* possible to envisage a permanent vascular system, but it would be very different from the one that evolved. The most obvious way of producing a complex system that can last indefinitely is to duplicate essential components. Then one can be repaired, whilst the other continues to function (see Note 2.1). Thus for the vascular system one would need two hearts and two sets of major arteries. Even so, each set would have to have much better means of self-renewal and repair than we see in extant vascular systems, and the same is true of the brain. Evolution falls very far short of producing *both* built-in redundancy *and* the means of replacing post-mitotic cells in essential organ systems.

The long-lived proteins of the lens and arteries have already been briefly discussed. Other very important proteins that are not turned over, or turn over very slowly, are those found in connective tissue. The collagen of connective tissue has been much studied in ageing research. In fact, collagen is a family of related proteins, in which the different polypeptide components associate to form distinct triple-helix linear molecules. Bundles of collagen molecules form fibres, which in turn form tendons and other types of connective tissue. Young collagen has a degree of elasticity, but with age this elasticity declines. It is known that this is due to the cross-linking involving the amino groups of lysine side chains. The chemistry of collagen cross-linking is complex and cannot be reviewed here in detail (see also the section on 'Protein modifications' in Chapter 4). Most gerontological investigators have simply measured by physical methods the changes in elasticity of collagen during ageing. This proves to be one of the best biomarkers of the ageing process, since the change can be quantitated with reasonable accuracy. For example, there have been many studies of the cross-linking of collagen in rat tail tendon fibres (see Fig. 2.1). The cross-linking increases linearly with age, but more slowly in animals that have an increased lifespan brought about by dietary restriction, or by removal of the pituitary (Everitt, Olsen & Burrows 1968; Everitt, Seedsman & Jones 1980). Although the chemistry of cross-linking in different species is very similar or the same, it is remarkable that the process occurs much faster in short-lived species than in long-lived ones. This is discussed

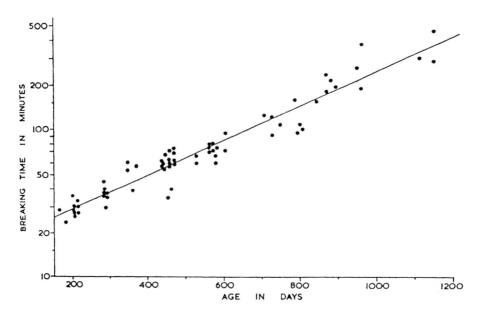

Figure 2.1. The increase in cross-linking of rat tail collagen fibres with age. The assay measures the breakage of fibres in 7 *M* urea, pH 7 at 40°, with a 2-g load. Compare with the chemical measurement of cross-links in collagen in Figure 7.4. (Reproduced with permission from Everitt, Olsen & Burrows, *J. Gerontol.* **23,** 333–6 [1968]. Copyright © The Gerontological Society of America.)

further in Chapter 7. In some cases collagen can be replaced, for example, in wound healing, but in many situations the structure of the organ or tissue in which it occurs makes replacement very difficult or impossible. In such locations collagen will undergo progressive chemical changes, including cross-linking, which will affect its normal functions and therefore the function of the tissue or organ.

Another protein of connective tissue that is very long lived is elastin, which again is in a family of related molecules. Although studied less extensively than collagen, it is very probable that progressive chemical changes in elastin result in a loss of its normal tissue functions during ageing. In fact, most proteins that have defined mechanical functions, collectively known as 'structural proteins', survive for long periods, and this means that they are liable to accumulate deleterious post-synthetic changes. Although there is much evidence that many defective proteins are recognised and degraded (see Chapter 3), in the case of long-lived structural proteins such a maintenance mechanism is often not available or not effective. Thus, there appears to be no mechanism specifically to remove highly cross-linked collagen or elastin and replace these molecules with new ones.

In machines, the moving mechanical parts are most subject to stress and wear, and essentially the same is true of joints in vertebrates. Prolonged use leads to degenerative joint disease. Although cartilage is active in metabolism, attempts by the organism to repair constant wear and tear are in the end unsuccessful, and one end result is osteoarthritis (see Chapter 8), which is a very common age-related disease. It is quite reasonable to compare the mechanical features of joints with those of machines. Almost always the moving parts of a machine that have suffered wear cannot be repaired, but can only be replaced. Essentially the same is true of joints, since surgical replacement with artificial structures is often successful.

So far in this chapter, most of the discussion has been concerned with long-lived molecules, cells and structures that cannot be replaced; but in the mammalian organism there are many dividing cells, and it might be thought that these are capable of continuous replacement. In fact, observations on dividing cells that occur in various parts of the body strongly suggest that such cells cannot divide indefinitely. The limited ability of cells and tissues to survive has often been examined by transplantation experiments. For example, it was found that transplanted rat mammary tissue would grow and differentiate normally in a recipient animal, and this tissue could again be successfully transplanted. However, after several sequential transfers, the ability of the donor cells to proliferate was lost (Daniel et al. 1968; Daniel & Young 1971). Similar results have been obtained when bone marrow cells were serially transferred in mice, or antibody-producing memory cells (Williamson & Askonas 1972; Ogden & Micklem 1976, but see Harrison 1979, 1984, 1985; Harrison, Astle & Stone 1989, and Note 2.2). In earlier experiments with inbred mice, whole pieces of skin were transferred from animal to animal. Although the skin tissue survived longer than the lifespan of the animal, it did not escape intrinsic senescent changes (Krohn 1962).

These studies of cell ageing *in vivo* have been complemented by the many studies of ageing *in vitro*. It has been shown with many types of diploid cell that the number of successive cell divisions is limited (reviewed in Chapter 5). It is a reasonable supposition that cell lineages *in vivo* also have a similar limit. Although it has never been directly demonstrated, it is at least a likely possibility that one cause of ageing is the loss of proliferative potential of cells required for the normal function of any tissue or organ, which depends on the replacement of cells from a stem line source. This is suggested by the fact that skin fibroblasts from elderly individuals have a shorter lifespan *in vitro* than those from young individuals. The skin tissue does not run out of a source of fibroblasts, but the number and growth potential of those remaining is reduced. This would affect the integrity of the skin tissue and the ability to repair wounds. It has been shown in experimental animals that the rate of wound healing declines with age (Bruce & Deamond 1991).

It is probable that whitening of the hair is due to the age-related loss of pigment-forming cells (melanocytes). If part of a mouse with pigmented hair is irradiated with a dose that kills dividing cells, then that area turns white because the pigment-forming cells are killed and not replaced (see Note 2.3). It is possible that a continual division of melanocytes is another example of the limited proliferative potential of somatic cells. Cells forming the hair itself are more resilient, although hair is also lost during ageing.

One feature of higher organisms is the remarkable stability of cell phenotypes. Lymphocytes divide to produce lymphocytes, fibroblasts produce fibroblasts, and stem line cells retain their determined state, that is, their ability to produce daughter cells that will differentiate into a specific cell type, or in some cases several specific cell types. There are clearly stable epigenetic controls that determine the spectrum of genes that are active in any specific cell and also those that are also inactive. The basis of these controls is not understood; however, it is evident that the controls can break down under certain circumstances. This happens in the development of cancer. Instead of a cell behaving as it should in a particular tissue, it loses control over the activity of genes, which in turn can trigger other events. If we consider the large number of cells in a mammal and the complexity of all the normal cellular controls, it is not surprising that epigenetic defects sometimes occur that can give rise to abnormal cell growth. Post-mortems of old animals or people usually reveal incipient or well-developed tumours, although they may not have been the cause of death. The question at issue here is the degree to which a complex animal can control its cells to prevent abnormal growth. It is clear that cancer in large, long-lived species is very much less common on a per-cell basis than it is in small short-lived ones (see Chapters 7 and 8). Thus it appears that controls can be improved with increased lifespan, but this is a relative or modulated change, and it is perhaps unlikely that an animal could completely prevent tumours arising during the course of its life.

The gerontologist Leonard Hayflick (1987) suggested that the correct question was not 'Why do we age?', but 'Why do we live as long as we do?' In this chapter, I have given some examples, but by no means a comprehensive list, of body components that inevitably lose function with age. In spite of the problems of maintaining structure, together with tissue and cell homeostasis, the larger mammalian species, as well as many fish, reptiles and birds, do live an extremely long time. Individual post-mitotic cells of the brain or heart can function for at least 100 years. We know that in the evolution of mammals, longevity increased (see Chapter 7), so the means of preserving cells, tissues and organs must also have improved. The ability of organisms to defy the laws of thermodynamics and survive for long periods of time depends completely on a wide range of repair and maintenance mechanisms, which are discussed in the next chapter.

3

Maintenance of the adult organism

The growth and development of the mammalian organism to the adult is normally followed by a reproductive phase, which comprises a considerable part of the whole lifespan. The adult organism is not in equilibrium or a steady state, but it retains all its physiological functions for a long period, which allows it to search for food, reproduce and care for its offspring. During this period, as well as during development, these functions are maintained by a variety of mechanisms. Some of the major maintenance mechanisms are described in this chapter.

DNA REPAIR

The DNA genome is the repository for all the genetic information of any species. It is essential that the integrity of this information be protected, both during the division of somatic cells and in its transmission from generation to generation. The intact DNA is subject to a wide variety of defects and damage, which are normally repaired. The study of DNA repair mechanisms was pioneered in bacteria and bacteriophages, but there is now much direct and indirect information about repair mechanisms in higher organisms as well (reviewed by Friedberg 1985; Sedgwick 1986). Repair processes can be broadly divided into those that act on so-called spontaneous changes in DNA, and those that act on DNA that has been damaged by radiation or chemical attack.

Spontaneous chemical changes in DNA are very frequent. Reactions that are particularly well known include the deamination of cytosine residues to uracil, the loss of purine residues from DNA (referred to as 'apurinic' or, more generally, 'abasic' sites), and the frequent deamination of 5-methyl cytosine to thymidine. It has been calculated that up to 10 000 purine residues are lost from a mammalian cell per day (Lindahl 1979, 1993). For about 10^{14} cells in the body, this works out at around 10^{13} purine residues lost per second in an adult human. As well as changes in bases, there can be single-strand breaks in DNA, and less commonly, double-strand breaks. It is probable that most of these changes are more likely in cells that are metabolically active than in those that are quiescent, and one reason for this is that oxygen

21

free radicals generated by respiration are one cause of 'spontaneous' DNA damage. It is also likely that base changes occur more frequently in the single-stranded DNA produced during transcription and DNA replication than they do in double-stranded DNA.

Enzyme systems exist to repair all this damage. Uracil is recognised by uracil glycosylase, which removes the incorrect base. An endonuclease widens the gap produced, and new DNA is synthesised to fill the gap. Abasic sites are repaired by the same mechanism. The deamination of 5-methyl cytosine to thymidine produces a normal DNA base, so a special repair mechanism is required that recognises a T–G mismatch, removes the offending T and substitutes the correct C (Wiebauer & Jiricny 1989, 1990). Single-strand breaks can be readily repaired, since the intact polynucleotide chain provides a template, but the repair of double-strand breaks depends on a special mechanism (see later in this section).

DNA can also be damaged by environmental agents. It is well known that DNA is the most important cellular target for various types of ionising radiation, and also for ultraviolet light (UV). Because UV does not penetrate beyond the outer layer of the skin, it produces primarily skin damage, with some immunological damage, but it can also strongly affect the lens and the retina of the eye. Photons of UV light are absorbed by DNA bases, which may then undergo chemical changes. One of the best-studied lesions in DNA is the dimer formed by the covalent joining of adjacent pyrimidine bases in DNA. This pyrimidine dimer is a substrate for a specific excision enzyme that cuts out the dimer, and the subsequent gap is widened and then refilled by other enzymes. Other lesions, such as the 6 : 4 pyrimidine derivative, are also produced by UV light and are repaired. Changes induced by ionising radiation are less specific: if the track of ionisation is very close to DNA, active oxygen free radicals can generate a single-strand break, and less often a double-strand break. In addition, there are various base adducts produced by this type of radiation.

There are many metabolites, produced by micro-organisms, that can damage DNA. For example, aflatoxin is produced by fungus that commonly contaminates peanuts. It reacts with DNA to produce a so-called bulky lesion, which can be repaired by the same excision repair process that acts on UV damage. In fact it is well known that mammalian cells that are never normally exposed to UV light are nevertheless able to repair this damage, as shown by the study of these cells in culture. This shows that environmental agents, such as aflatoxin, produce DNA damage that is a bulky lesion sufficiently similar to that produced by UV light to be repaired by the same enzymes.

Mitomycin C is an antibiotic that inhibits DNA synthesis by cross-linking of strands. Cross-linking, as well as the formation of double-strand breaks in DNA, poses a problem for the organism, because there is no intact template that the repair enzymes can utilise. It is now known that double-strand breaks in bacteria and yeast are repaired by a special mechanism involving genetic recombination. If the cell has a homologous undamaged stretch of DNA in the same nucleus (as is the case in the G2 stage of the cell cycle in a haploid cell, or in any diploid cell), then this can be used as a template to repair the damage. Double-strand breaks are also repaired in mammalian cells, but it is not established that this requires genetic recombination. It is known, however, that DNA-damaging agents can greatly increase the frequency of recombination or crossing-over of sister chromatids. (Sister chromatid exchanges [SCEs] can be detected by staining procedures that distinguish the chromatids at metaphase.) Mitomycin C treatment greatly increases the frequency of SCEs, and is also known to induce recombination between homologous chromosomes.

Many reactive chemicals fall into the class collectively known as alkylating agents. These chemically modify DNA bases, and a particularly important product is O^6-methyl guanine. This is removed by a surprising reaction, namely, the transfer of the methyl group to a protein known as methyl transferase. The protein is not an enzyme, since it cannot act again after it is methylated. It is a suicide protein, which can only be degraded after it has become methylated. This reaction well illustrates the vital importance of DNA repair, since in this case a whole protein molecule is synthesised to remove the chemical modification of a single base. A major repair mechanism in bacteria is known as the 'induced SOS response'. Damage to DNA, induced for instance by UV light or mitomycin C, results in the synthesis of a group of proteins that are required for SOS repair. This repair is particularly important when other mechanisms fail, and it involves an 'error-prone' process. In the SOS response, the DNA polymerase can replicate a damaged template, which under normal circumstances does not happen, and it is then likely to insert incorrect bases. The induced SOS response is therefore a major source of induced mutation. It is likely, but not yet proved, that a similar error-prone repair mechanism exists in mammalian cells.

Many mutants of mammalian cells are known (as well, of course, as those in microbial species) that are defective in the repair of DNA damage. These mutants may be much more readily killed by a particular DNA-damaging agent, whereas in others there is cross-sensitivity to several different kinds of such agents. Individuals that inherit a mutation resulting in defective DNA repair may have multiple abnormalities, including chromosome breaks or rearrangements, an increased incidence of tumours, neurological defects and an impaired immune response. This demonstrates that DNA repair is a very important maintenance mechanism that is vital for normal cell survival and function. Where damage for one reason or another is not repaired, or is imperfectly repaired, the resulting abnormalities in DNA may contribute to the ageing process (see Chapter 4). Repair depends on the efficiency of many enzymes, and of course this depends on information in DNA. Since repair mechanisms are genetically determined, we might expect greater or less investment in DNA repair in different species. In fact, there is very good evidence that long-lived species are more efficient in repair than short-lived ones, which is reviewed in Chapter 7.

ACCURACY IN THE SYNTHESIS
OF MACROMOLECULES

In the synthesis of the polymeric macromolecules DNA, RNA and proteins, accuracy in assembly is essential; otherwise their functions would be seriously impaired. However, the enzymes that are necessary for the synthesis of these macromolecules are not by themselves completely accurate. As well as acting on their normal substrates, they often also have some activity on related substrates or analogues. This creates a problem in the accurate synthesis of macromolecules, which, with regard to protein synthesis, was first pointed out by Pauling (1958). From his knowledge of basic chemical principles, he estimated that an enzyme could not be expected to distinguish between two related amino acids, such as isoleucine and valine, with a specificity greater than about 99%. Thus, for every correct insertion of an amino acid in a polypeptide, one might expect an incorrect insertion of a related amino acid at a frequency of about 1%. Yet we know that the accuracy of protein synthesis is much greater than this (see later in this section).

The problem is most formidable in the case of DNA, where we have very good estimates of the accuracy of synthesis. It is vital for the organism to transmit from cell to cell and from generation to generation faithful copies of the genetic information, and in practice it has been demonstrated in a variety of organisms that the error level, or spontaneous mutation rate, is in the range of $10^{-8}-10^{-10}$ base per cell division (Drake 1970). Complementary base pairing alone certainly cannot achieve that. After the elucidation of the structure of the DNA double helix, Watson and Crick speculated that spontaneous mutations might occur by the tautomeric shift of bases, which would then result in the formation of hydrogen bonds with a non-complementary base. It was very soon apparent, however, that this would produce mutations at a rate very much greater than is actually observed. The reasons for the accuracy of DNA replication are now to a large extent understood from the many studies of bacteria and bacteriophages, and it is likely the same or similar mechanisms are used in higher organisms (reviewed by Kornberg & Baker 1992). DNA replication in *Escherichia coli* is carried out by the DNA polymerase III (Pol III)holoenzyme, which is a complex of many proteins. The basic specificity is achieved by the enzyme inserting complementary bases in the growing chain, but this specificity is not particularly high, so the incorrect insertion of a wrong base is quite common. One of the subunits of the enzyme has the ability to recognise a mispaired base, and the enzyme complex also has exonuclease activity. For every hundred or thousand steps forward, the enzyme takes one step back and removes an incorrect base. This process is called 'proof-reading'. An enzyme lacking the ability to recognise a mispaired base (because it has a mutation in the subunit required for recognition) has a mutator phenotype, which means that many spontaneous mutations occur during chromosome replication.

Two general principles have emerged from the studies of the proof-reading process. First, proof-reading requires energy. The incorrect nucleotide is initially a triphosphate, but after it is removed it is a monophosphate; thus the energy of two phosphate bonds is used up in every proof-reading step. Second, there is a strong relationship between the accuracy of a polymerising enzyme and the speed at which it operates: the greater the speed, the lower the accuracy (see Note 3.1). If the role of DNA polymerase were simply to replicate chromosomal DNA as fast as possible, it would not carefully discriminate between bases; it would simply keep going and produce many mutations. Proof-reading depends on slowing down the rate, to ensure complementary base pairing and also to go backwards to correct errors when necessary. There is no theoretical reason why an enzyme complex could not have evolved that carried out more than one proof-reading step. Such an enzyme would have greater accuracy, use more energy and replicate DNA more slowly. In the evolution of the organism an optimum level of accuracy is achieved, in which there is a balance between accuracy and rate of replication, but this need not be the same for all species.

The proof-reading of the Pol III enzyme complex in *E. coli* does not, however, achieve the accuracy of DNA synthesis that can be demonstrated in many genetic experiments where the mutation frequency is accurately measured. It is now known that there is a backup process, usually referred to as mismatch repair. If the polymerase proof-reading mechanism fails to detect a mismatched base pair at the point of replication, then the DNA in the nascent growing chain contains a non-complementary base pair. There is an enzyme system that detects this and restores the correct base sequence (reviewed by Modrich 1991). To do this, the system must be able to distinguish the new incorrect base from the old correct base in the DNA duplex. This depends on the post-synthetic methylation of adenine in a short four-base sequence in DNA (GATC). Methylation is carried out by an enzyme known as the Dam methylase, which acts on the newly-synthesised DNA chain. This means that there is always a short region of DNA where one chain is the methylated parental template and the other is the nascent, unmethylated DNA. The mismatch repair system monitors the nascent chain and repairs any incorrect or mismatched bases. Mutations that inactivate the dam methylase, or any of four other enzyme components of the mismatch correction system, produce a mutator phenotype.

To ensure accuracy in DNA replication the organism must therefore synthesise all the components of the replication complex and also those of the mismatch repair system. Moreover, this repair depends on the excision of a portion of the new DNA polynucleotide chain and the resynthesis of the correct sequence. All these events require the expenditure of energy by the organism. In higher organisms it is known from genetic studies that the replication of DNA is at least as accurate, or perhaps more accurate, than in bacteria. It is not yet known how this accuracy is achieved, but it is a safe assumption that expensive proof-reading and/or mismatch repair mechanisms are essential.

RNA polymerases required for transcription are also complex enzymes, and they incorporate a proof-reading step (Libby et al. 1989; Libby & Gallant 1991). However, the product of transcription is single stranded, so the equivalent of a mismatch repair process is not possible. It is not surprising, therefore, that RNA synthesis is much less accurate than DNA synthesis. There are far fewer determinations of the accuracy of RNA synthesis *in vivo* than there are of the accuracy of DNA synthesis, but from those that exist, it is likely that there is one error in about 10^4–10^5 bases. More information is available about errors in the replication of viruses with RNA genomes, and it is clear that the mutation frequency is several orders of magnitude higher than mutation in DNA genomes. A small RNA virus with 104 nucleotides may have on average about one spontaneous mutation in its genome, whereas a corresponding DNA virus would have 10^4–10^5 fewer mutations.

As well as transcriptional accuracy, higher organisms must also splice transcripts to produce messenger RNA (mRNA). To preserve the coding information, including the reading frame, it is essential that splicing be accurate and lead to the joining of adjacent exons without the insertion or deletion of a single RNA base. How this is achieved is unknown; the action is carried out by a multicomponent complex, the SnRNP particle or 'snurp', which contains both RNA and proteins. It is possible that the specificity is achieved by complementary RNA base pairing, using a mechanism related to the well-known genetic recombination of DNA molecules (see Note 3.2).

Having achieved accurate transcription and splicing, and additional processing of the mRNA, transport of the completed message to the cytoplasm now allows the translational assembly of the appropriate polypeptide chain to proceed. This is a complex metabolic process utilizing ribosomes that consist of two major subunits, each composed of ribosomal RNA (rRNA) and a total of more than 50 types of protein. It is known that the ribosome plays a positive role in the accuracy of translation, as well as in peptide bond synthesis, since mutations in ribosomal proteins can alter accuracy. It is also necessary for individual amino acids to be specifically attached to the appropriate transfer RNA (tRNA) molecules. The anticodon of the tRNA with the correct amino acid attached must pair correctly with the appropriate triplet of mRNA, so that the information in the RNA is translated into an error-free polypeptide. There are many possible sources of error in this whole assembly process, so it is not surprising that proof-reading is also involved in protein synthesis. An inappropriate aminoacyl tRNA can be recognised and discarded. The enzyme kinetics of proof-reading in protein synthesis were first worked out on theoretical grounds, and later confirmed by experiment (see Note 3.1). An additional mechanism known as 'ribosomal editing' has also been proposed by Menninger (1977). This is a process whereby peptides are dissociated from the ribosome, if a peptidyl–tRNA linkage is formed at an inappropriate codon in the message.

There has been much interest in the relationship between the accuracy of protein synthesis and ageing (see Chapter 4), so it is surprising that there are still relatively little data on the actual accuracy *in vivo*, especially in mammalian cells. In bacteria, the best estimates suggest an overall error level in translation of 10^{-3}–10^{-4}, perhaps about ten times higher than transcriptional errors (Kirkwood, Holliday & Rosenberger 1984, and see Note 3.3). In protein synthesis the same principle applies as for DNA synthesis, namely, that the rate of formation of a polypeptide is inversely related to accuracy. Mutations in ribosomal proteins that slow down chain elongation can increase accuracy, and other mutations reduce accuracy. One can expect an optimum in which accuracy and rate of protein synthesis are balanced, and this optimum may well vary in different species, or in particular biological contexts.

In conclusion, we see that the assembly of macromolecules depends on a series of complex reactions involving enzymes, or other proteins, as well as RNA. Obviously the synthesis of DNA, RNA and proteins involves a considerable proportion of the total metabolic resources of every cell, and of this proportion a substantial part ensures accuracy of synthesis. This includes not only the energy actually involved in the proof-reading, but also the need for ribosomes, snurps, polymerase complexes, mismatch repair enzymes and so on. It must also be emphasised that the accuracy of macromolecular synthesis is under genetic control. All the components required are coded for by genes, and it is possible to envisage the evolution of highly accurate polymerisation mechanisms, or much less accurate ones, depending on the species of organism, or type of cell. As we shall see in Chapter 7, it may well be the case that germ line cells are more accurate in macromolecular synthesis than somatic cells.

DEFENCE AGAINST OXYGEN FREE RADICALS

Free radicals are chemical species that have an unpaired electron in an outer orbital. The radical is extremely reactive, but also very unstable. In the context of biological damage, oxygen or oxidative free radicals are the most important, and discussion of 'free radical' damage normally refers to oxygen radicals. These can react with inorganic or organic molecules including proteins, lipids, carbohydrates and nucleic acids. Oxidative reactions in mitochondria are a major source of oxygen free radicals. In the normal respiratory pathway oxygen is reduced to water, but in this process partially-reduced oxygen species are produced, the most important being superoxide, O^-_2, hydrogen peroxide, H_2O_2, and hydroxyl ions, OH^-. These radicals are also produced by oxidative enzymes in the endoplasmic reticulum and elsewhere. P450 cytochromes have an essential role in removing toxic chemicals (see the section on 'Toxic chemicals and detoxification'), but they also generate oxygen free radicals. The peroxisome organelle degrades fatty acids and other molecules, and produces hydrogen peroxide as a by-product. Phagocyte cells, especially macrophages, destroy cells infected with bacteria or viruses with an oxidative burst of superoxide and hydrogen peroxide, as well as the radicals nitric oxide (NO) and hypochlorite (OCl⁻). Chronic infections result in excessive phagocyte activity, chronic inflammation and other side effects. Lipid auto-oxidation in the gastrointestinal tract produces hydrogen peroxide from which hydroxyl radicals are derived. External agents such as ionising radiation or ultraviolet light also generate oxygen free radicals in tissues.

Three major enzymes are involved in the removal of oxygen free radicals. These are superoxide dismutase (SOD), catalase and glutathione peroxidase, and the pathways involved are shown in Figure 3.1. It should be noted that damaging hydroxyl radicals are produced from peroxide by the Fenton reaction mediated by transition metal ions, especially iron (Fe^{++}). They can also be formed by the reaction of hydrogen peroxide with superoxide. As well as the enzymic defence against free radical damage, there are also a variety of antioxidants that, by reacting with oxygen free radicals, nullify their effects (Sies 1993). These include α-tocopherol (vitamin E), ascorbic acid (vitamin C), uric acid, sulphydryl containing compounds such as cysteine and glutathione, bilirubin, ubiquinol and carnosine. Oxygen radicals are produced by photosynthesis in plants and chloroplasts contain large amounts of β-carotene, which is an active sink for radicals. Carotene is a normal component of diet, and it is likely that it helps remove active radicals in animals. The protein transferrin may act as an antioxidant by removing free iron, which can otherwise mediate hydroxyl radical formation.

Damage to DNA includes the formation of several base adducts, which are excised by repair enzymes and eventually excreted in urine (see Note 3.4). This has been exploited in important recent studies by Ames and his collaborators (Ames et al. 1985; Adelman, Saul & Ames 1988; Richter, Park & Ames 1988; Fraga et al. 1990; Wagner, Hu & Ames 1992; Ames, Shigenaga & Hagen 1993). Using sensitive chromatographic techniques they measured the quantities excreted by animals with different metabolic rates, and found that the amount of adduct per kilogram of body weight was proportional to metabolic rate (see also Chapter 7). This is to be expected if radicals are generated primarily by respiration, but the results also show that DNA is a real target for damage. The same group has also demonstrated that damage in mitochondrial DNA is very much greater than damage to nuclear DNA, as would be expected, since the respiratory chain in mitochondria is a major source of free radicals. In fact, a fundamental defence against oxygen free radical attack may be the localisation of aerobic respiration to the mitochondria. By this means the oxygen free radicals that are generated do not often gain access to chromosomal DNA, which is therefore protected in a relatively oxygen free environment (Lindahl 1993).

Oxygen free radicals also produce lipid peroxidation of membranes of cells and organelles. This is likely to be particularly important in long-lived, non-dividing cells such as neurones. These cells are also very active in metabolism, and there is a high lipid content in the myelin sheath of axons. Lipid

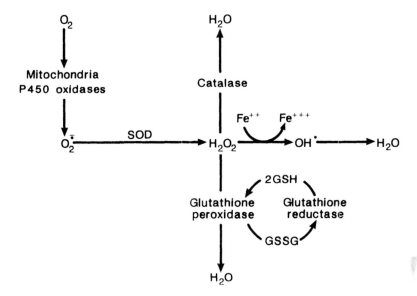

Figure 3.1. The major pathways for the formation and removal of oxygen free radicals.

peroxidation is believed to be one of the sources of the so-called age pigment lipofuscin, which accumulates in brain and other tissues with long-lived cells, such as the heart. The chemistry of lipofuscin formation is not understood, but since the material accumulates it must be the by-product of reactions that cannot be prevented or reversed by the normal cell maintenance process. It is probable that oxidation of proteins by radicals is a continuous, ongoing process in cells, as is the removal of the abnormal molecules (Davies & Goldberg 1987). Oxidation of proteins is one of many abnormal post-synthetic changes that can occur (see Chapter 4). Organisms have important defences that can recognise and remove such molecules (see the following section).

REMOVAL OF DEFECTIVE PROTEINS

Proteins are subject to a wide variety of chemical modifications in the cell. Many of these are an integral part of protein function, for example, specific phosphorylation, glycosylation or methylation. These modifications are essential for important regulatory processes, in membrane function and in many other enzymic or structural contexts. We are concerned here with abnormal post-synthetic modification of proteins. The amino acid side chains of proteins can be altered in many ways. Deamidation occurs in asparagine and glutamine residues, and the terminal amino group of lysine reacts with various sugars (collectively known as non-enzymic protein glycation or glycosylation). Lysine residues are also involved in the formation of cross-links in proteins such as collagen. Proteins may also become phosphorylated or methylated in abnormal positions. Recently a class of proteins known as chaperonins has been discovered; these have an important role in protein folding, and also in their transport to particular locations (Ellis & van der Vies 1991; Gething & Sambrook 1992). If the correctly-folded molecule is at a higher energy state than one that has not been folded, then it may revert to a non-functional, low-energy state. More generally, proteins are subject to partial or complete denaturation.

It is not surprising therefore that all cells are equipped with a battery of proteases that are capable of removing abnormal molecules (see Makrides 1983; Dice 1987; Hipkiss 1989). One important pathway involves a small protein known as ubiquitin (Rechsteiner 1987, 1991; Hershko & Ciechanover 1992). Ubiquitin appears to be capable of recognizing abnormal molecules and becomes attached to them, by a mechanism that is not clearly understood. However, once the protein is tagged with the ubiquitin it becomes a substrate for proteases, which reduce it to its constituent amino acids. Some proteases require a source of energy, ATP, in order to degrade their substrate. Others are able to cleave only specific peptide bonds, and the resulting peptides become substrates for other enzymes.

Proteins, of course, have very different half-lives in the cell. Some, such as collagen, elastin and crystallin, are very long-lived, and individual molecules may never be replaced during the lifetime of the organism. Others are very short-lived, and there may be a specific reason for their removal if they have a specific regulatory role, for example, in the control of the cell cycle. Many proteins, such as housekeeping enzymes, would normally be expected to survive as long as they are functional. Denaturation, or one of the other abnormalities mentioned, signals their demise at the hand of proteases. It was demonstrated by Robinson, McKerron and Cary (1970) that the rate of spontaneous deamidation of asparaginyl and glutaminyl residues in peptides and proteins is a function of neighbouring amino acids in the polypeptide chain. It is therefore possible that a given protein could evolve to a stable or unstable form, depending on the amino acid sequence. Such mutations need not necessarily affect the function or specificity of the enzyme, but might be related to the lifespan of cells or organisms, although direct evidence for this is lacking.

The importance of protein turnover in animals is demonstrated by the need for a continual source of nitrogen, including amino acids, in the diet, and also by the continual secretion of nitrogen in the form of urea in mammals, or uric acid in some other vertebrates. This continual nitrogen metabolism in the body is in large part due to the synthesis and subsequent breakdown of proteins. In the degradation of proteins by proteases, some amino acids are reutilised, but many are catabolised and their nitrogen is excreted. The enormous amount of energy expended on protein anabolism and catabolism is in large part a process of maintenance, to ensure normal cell structure and function. Without such a process cells would very soon get clogged up with denatured or other abnormal protein molecules.

THE IMMUNE RESPONSE

Slow-breeding complex organisms are continuously exposed to infection by rapidly-growing pathogens or parasites, and any intrinsic resistance may be readily overcome by mutations to virulence by the invading organism. As a consequence, vertebrates have evolved the very complex defence mechanism of the immune response. Immunology is, of course, a science in its own right, and there would be little point in attempting here to review in detail the vast amount of information that is now available (but see Roitt 1988). Instead, only a brief outline of the essential features of the immune defence mechanisms is included. It is, however, essential to emphasise the fact that the complex cell and tissue components of the immune system comprise *in toto* a major maintenance mechanism in mammals, birds and other vertebrates. The effect of the failure of the immune system is graphically illustrated by the fate of patients suffering from AIDS, who are infected by an HIV virus that destroys an essential component of the immune system. These patients usually die from pneumonia or other infections, and they also develop characteristic types of tumour, particularly Karposi's sarcoma.

The striking feature of the immune response is that it is based on several lines of defence. It is convenient to divide these into four major components, but they are by no means isolated, as there are many important interactions among them. The two major lines of defence in the immune system are the intrinsic, innate or non-specific mechanisms, and the acquired, adaptive or specific mechanisms. Within each of these lines there are humoral, or noncellular, mechanisms and cellular mechanisms. The major cellular component of the intrinsic defence mechanisms is the macrophage, but neutrophils and natural killer (NK) cells are also involved. It has long been known that macrophages are capable of engulfing and killing invading micro-organisms, and an important component of the killing process is dependent on the production of oxygen free radicals. Neutrophils also have antimicrobial activities, and NK cells can reorganise and lyse virus-containing cells, without prior sensitisation. The major intrinsic humoral defence mechanism is the complement system, which consists of about twenty proteins and their cleavage products. There are components that are activated, for example, by bacterial polysaccharides, and components that are effectors. These have a role in increasing vascular permeability and inflammation, in chemotaxis and bacterial lysis, and also in the sensitisation (opsonisation) of invading bacteria prior to phagocytosis by macrophages.

The acquired immune response is dependent on T and B lymphocytes. A major feature of the response is the distinction that is made between self and foreign antigens. In brief, all lymphocytes that would react with self antigens are eliminated during development. The remaining lymphocytes contain a huge repertoire of immunoglobulin molecules, potentially capable of recognising the foreign antigens of invading organisms (including non-self tissue; see below). The humoral response is based on the production of soluble antibodies, which may neutralise bacterial or other toxins, or virus particles. The B lymphocytes capable of producing the correct antibody may be present in small numbers, so a mechanism exists for the clonal expansion of these cells when a particular antigen is encountered. Thus, the organism amplifies its defence and also acquires an immunological memory that enables it to deal more effectively with any subsequent infections. There is also a selection of cells producing antibodies with greater and greater affinity for a given foreign antigen. The ability of the organism to generate very large numbers of immunoglobulin molecules depends on special genetic mechanisms. There are many alternative germ line DNA coding sequences combined by DNA splicing mechanisms into a single coding sequence in individual lymphocytes. In addition, there are frequent somatic mutations in the variable regions of immunoglobulin, which bring about genetic diversity.

The acquired cellular immune response is dependent on T lymphocytes. Whereas B lymphocytes secrete immunoglobulin, T cells retain them at their surface, and they recognise foreign antigens by contact. T cells produce a series of cytokines or growth factors, which activate other cells of the immune system, and are also responsible for the distinction between self and non-self human leukocyte antigens. These are present in the surface of most cells, and since the human leukocyte antigen (HLA) gene cluster is highly polymorphic, most transplanted tissues express HLA antigens that are foreign. These cells are therefore recognised and rejected. This is not, of course, a biological function of HLA antigens; instead, it is now evident that HLA molecules have a role in presenting foreign peptide antigens on the surface of cells, for example, a peptide derived from an infecting virus particle. The recognition of this antigen by a T cell triggers the destruction of the virus-infected cell and therefore prevents the spread of the virus. T cell populations have a very large repertoire of surface immunoglobulins, which means that cells specific for a foreign antigen may be a minute fraction of the total. T lymphocytes, like B lymphocytes, are also capable of clonal proliferation when activated. It is notable that some of the most successful pathogenic viruses have RNA genomes, such as the HIV retrovirus and influenza. As mentioned above (see 'Accuracy in the synthesis of macromolecules'), there is a very high mutation rate in such genomes, so the virus is continually changing the amino acid sequence of its proteins (antigens) and thereby often preventing the host from mounting a successful long-term response, as well as making it very difficult for immunologists to develop effective vaccines.

The immune system depends on bone marrow tissues, the thymus in young animals, the lymphatic glands, and large numbers of cells in blood and lymph. Together it is a very complex organ system, characterised by multiple interacting lines of defence. The defence is mounted primarily against invading pathogens, although it is widely maintained that it is also essential for immunological surveyance and the detection and destruction of abnormal tumour cells. Such cells do produce new antigens, such as embryonic antigens that are absent in adult tissue, but their significance in eliciting responses from the immune system is still a matter of controversy. The widespread use of drugs that suppress the immune response in transplant patients should clarify this issue, since the probability of these patients developing tumours, in comparison to untreated individuals, is continually being monitored. Be that as it may, it is abundantly clear that the immune system evolved to ensure maintenance and survival in a hostile natural environment that contains rapidly-reproducing pathogens.

TOXIC CHEMICALS AND DETOXIFICATION

One of the main defences plants have against herbivorous animals is the production of toxic chemicals in the leaves or other parts of the plants. It is now clear that a considerable proportion of plant metabolism (collectively known as 'secondary metabolism') is devoted to the synthesis of these chemicals. A large range of toxins exist, some directed against insects, some against mammalian herbivores, and so on. In response, animals have evolved specific detoxification mechanisms over a long period of evolutionary time. Thus, there is an ongoing conflict, or 'arms race', between plants and animals that has lead to the production in animals of a very sophisticated and complex detoxification system.

This is based primarily on a series of monooxygenase enzymes, known as the P450 cytochromes, coded for by the P450 gene superfamily (Lu & West 1979; Nebert & Gonzalez 1987; Parke, Ioannides & Lewis 1991). These can carry out a wide range of chemical reactions, including aliphatic oxidation; aromatic hydroxylation; oxidative deamination; nitrogen, oxygen and sulphur dealkylation; sulphoxide formation; nitrogen oxidation and hydroxylation, and oxidative dehalogenisation. The liver is a major site for detoxification, and the enzymes are primarily located in the endoplasmic reticulum of hepatocytes (the liver microsome fraction). However, other cells also contain P450 cytochromes. As well as the endogenous level of enzymes, there are also complex regulatory mechanisms that induce enzyme synthesis in response to an absorbed toxic compound. In the detoxification process, active intermediates are often formed that are rapidly converted to inactive derivatives and subsequently secreted. Much research in the field arises from the fact that man-made chemicals are also subject to the detoxification process, and the reactive intermediates may also be carcinogens or mutagens. Thus, for example, hydrocarbon carcinogens such as dimethyl benzanthracene (DMBA) or benzpyrene only react with DNA after they are activated to epoxides. It is not surprising that the enzymes that evolved to deal with plant toxins are sometimes less able to remove chemicals that appeared in our diet (or from other sources) very' recently. In the famous series of Ames tests, a very large number of chemicals were screened for mutagenic activity in bacteria by incorporating liver microsomes in the assay procedure. Evidence suggests that different mammalian species vary in their ability to detoxify chemicals. Studies with DMBA indicate that binding to DNA and production of mutagenic derivatives is greater in short-lived mammals than in long-lived ones (see Chapter 7). This indicates that the ability to deal with potential toxins is more effective in the long-lived species, possibly by the more rapid removal of an active intermediate, but much more information is needed to demonstrate this.

Detoxification of chemicals in the diet is a continual maintenance process that is essential for survival. Plant foods that are completely palatable nevertheless contain chemicals that are eliminated from the ingesting organism. Without the P450 enzymes, such foods would in many cases be highly toxic.

WOUND HEALING

One of the most obvious maintenance mechanisms in animals is their ability to repair or heal external or internal mechanical injury. We saw in Chapter 1 that some simple animals are able to regenerate the whole organism from a pool of totipotent cells. As animal evolution proceeded, this regenerative ability in general regressed, although we do see some remarkable powers of regeneration in vertebrates. For example, if the leg of the amphibian axolotl (genus *Ambystoma*) is severed, a perfect new limb is formed. Similarly cutting the optic nerve of a young amphibian, such as *Xenopus,* is followed not only by regrowth of the nerves, but also a complete matching of the thousands of axons from the retina to the corresponding positions in the tectum of the brain. Such regenerative ability has clearly been lost in mammals and birds.

Amphibian regeneration is an extreme case of wound healing, which has many relationships to development itself. The severed limb forms a blastema of cells that are akin to embryonic cells, and the morphogenetic processes they bring about are probably almost exactly the same as those that occur in normal limb development. It is not certain whether the blastema arises from a quiescent pool of cells, with the same or similar totipotent capacity as embryonic cells, or whether it is formed by 'dedifferentiation' of pre-existing differentiated or partially-differentiated cells in the undamaged limb.

Although regeneration in animals and birds is less than in lower vertebrates, the wound healing that is seen follows similar morphogenetic principles. When tissue is damaged the cells are stimulated to divide to repair the damage, and they in some way 'know' when the repair is complete. The study and understanding of the healing of wounds is a very important component of medical practice, and the details cannot be discussed here. In brief, the damage of tissue such as skin is followed by a series of events, which include inflammation, the stimulation of epithelial cells and fibroblasts to divide, the formation of blood vessels or capillaries (depending on the size of the wound) and the deposition of collagen, elastin, laminin, proteoglycans and other components of connective tissue. Macrophages are active in inflammation and remove infecting bacteria, as well as debris from the damaged tissue. The stimulus for cells to divide is the release of mitogenic growth factors. The treating of small wounds leaves no trace, but in more severe wounding the damage is not completely repaired. Instead, the tissue contracts and the remaining space is filled by scar tissue, which is imperfect in comparison to the original: it may have an incomplete blood supply, lack innervation, and in the case of the skin, it may lack normal hair follicles. Although much information exists about the events and cells involved in wound healing, the repair that is seen in skin, muscle, cartilage, bone or other tissues involves morphogenetic processes that are as yet not fully understood.

A model for wound healing can be seen in skin fibroblasts growing *in vitro*. Such cells divide in appropriate medium to form a confluent monolayer, after which the cells stop dividing owing to contact inhibition. If the monolayer is damaged by scraping a needle across it, then the cells at the edge of the artificial wound start dividing, the space is filled with a new monolayer and division ceases. As in the case of *in vivo* healing of minor wounds, no trace of the damage remains.

One very well understood example of repair is blood clotting. The damage to a vein or artery is followed by a cascade of events leading to the formation of a clot of blood. This is a gel containing a network of fibrin derived from fibrinogen, which is a soluble serum protein. The clotting reaction depends on platelet aggregation, calcium ions and the conversion of prothrombin to thrombin, which converts fibrinogen to fibrin. In patients with the disease haemophilia, one step in the cascade of reactions is defective, and blood does not clot. The disastrous effects of this are seen not only in the continuous bleeding of open wounds, but even more important, in the internal bleeding in joints and other locations subject to mechanical stress.

Wound healing is clearly essential for the repair of damage sustained in childhood and adulthood. It is, however, important to note that during the evolution of mammals some balance has been reached between the retention of efficient healing of minor wounds, and the loss of regenerative repair of major damage, such as the replacement of lost digits, the growth of severed nerves and so on. This balance is a good example of an optimum investment in maintenance. It would be selectively disadvantageous not to have the capacity to repair skin, muscle, vascular supply and so on, but it would also be selectively disadvantageous to invest major resources in the much greater capacity for repair that would be needed to replace brain tissue, sense organs, major nerves, limbs or digits.

Table 3.1. *Formal representation of gene expression in specialised cells*

	Proteins[a]						
	A	B	C	D	E	F	G
Cell type 1	√	√	√	√	—	—	—
Cell type 2	√	√	—	—	√	√	√
Cell type 3	√	√	—	√	√	—	—
Ectopic expression in cell type 3	√	√	—	√	√	√	—

[a]A and B are 'housekeeping' proteins, C–G are 'luxury' proteins.

STABILITY OF DIFFERENTIATED CELLS

It is well known that differentiated cells do not switch their phenotypes. This is fairly obvious for non-dividing cells such as neurones, but is it also true for dividing cells. A single fibroblast can produce a very large clone or population of cells all of which retain the same phenotype, and the same is true of T lymphocytes, endothelial cells, glial cells and so on. It would be an astonishing event if we saw a macrophage or myotube amongst a clonal population of fibroblasts. It is safe to conclude that such events either do not occur at all or occur with extreme rarity.

For convenience proteins can be divided into two general classes known as 'housekeeping' and 'luxury' proteins. Housekeeping proteins are present in all cells active in metabolism, where they mediate the vast number of biochemical reactions necessary for normal metabolism and viability. Luxury proteins, on the other hand, are produced by particular types of differentiated cell. Thus haemoglobin is a luxury protein in erythrocytes, and keratin in skin epidermal cells. The combination of proteins present in cells is formally represented in Table 3.1, where A and B represent housekeeping proteins and C to G luxury proteins. The particular phenotypes of the cells depend on the luxury proteins that are present, and different combinations of luxury proteins are possible. Cell type 3, for example, may have both some specialised proteins of 1 and 2, but not others. A rare aberrant event would be the ectopic synthesis of one of these (F). The table illustrates the basic observations, but it is an oversimplification in the sense that housekeeping enzymes may occur in different forms, known as isozymes, which are a characteristic of particular cell types. Also, it is known that alternative splicing of transcripts can produce a family of proteins, with similar or related functions, that are cell-type specific (Breitbart, Andreadis & Nadal-Ginard 1987; Smith, Patton & Nadal-Ginard 1989).

The problem for complex organisms is to stably maintain the phenotype of specialised cells, allowing for instance the synthesis of protein C in cell type 1, whilst preventing it in cell type 2. These stable controls are often referred

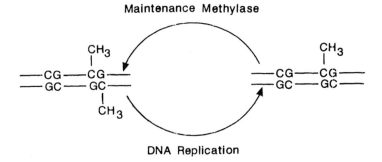

Figure 3.2. The modified base 5-methyl cytosine occurs in CpG doublets, but some are not methylated. Hemimethylated DNA is a substrate for a maintenance DNA methylase, which does not act on non-methylated CpGs. Thus, any given pattern of DNA methylation is heritable.

to as epigenetic, and it is generally agreed that they depend on specific DNA–protein interactions. It is commonly believed that all that is required in a specialised cell is a repertoire of transcriptional activators or repressors, which simply lock the cell into the desired phenotype. This may very well be the case, but the real problem is the *stability* of the controls required. This can be well illustrated by a particular example of epigenetic control, namely the existence of both an active X chromosome and an inactive X chromosome in the cells of a female mammal. It is well known that these chromosomes stably co-exist in a common nucleoplasm, without switching their activities (Gartler & Riggs 1983). If the control is dependent only on the presence of a particular set of regulatory proteins in the cell, then these differences between the X chromosomes would not be seen, or maintained. What is required is heritability of active or inactive states of genes at the chromosome level. It has been proposed that such heritability could be based on specific co-operative binding of proteins to DNA (Weintraub 1985), which once set up would be stably maintained through cell division. This remains a possibility, but such a mechanism of gene control has yet to be described. Another possibility, which is not yet widely accepted, is that the major controls depend on the covalent modification of bases in DNA, specifically the formation of 5-methyl cytosine at particular sites in DNA. There is good evidence that the pattern of methylation in DNA is inherited through the activity of a maintenance methylase (Fig. 3.2 and Note 3.5). Methylated DNA replicates to produce DNA in which one strand is methylated and the other is not, and the methylase acts on the unmethylated strand. This enzyme does not act on non-methylated DNA *in vivo*. Thus, a given sequence can be stably maintained in a methylated or non-methylated state through repeated cell division. In the case of the X chromosome, it is known that methylation plays a role in the silencing of the inactive X. The CpG islands of housekeeping genes are methylated on the inactive

X and unmethylated on the active X (see Note 3.6). These and many other observations show that methylation is an important component in the perpetuation of gene activity, or inactivity, of particular genes, and this 'locking in' of a pattern of gene activities may be responsible for the different phenotypes of cells (see Table 3.1). Whether DNA methylation is responsible for the switching processes that are necessary for the initial formation of different cell types is much more controversial, and only time will tell if this is or is not the case.

Evidence reviewed in later chapters suggests that epigenetic controls are more stably maintained in large long-lived animals than in small short-lived ones. If this is the case, then it is highly probable that more resources are used in the former to ensure the stability of cell phenotypes and to avoid abnormal heritable changes, which could lead to the emergence of tumours. Although we are very ignorant of the actual mechanisms, it is a reasonable assumption that maintenance of cell phenotypes is an extremely important process, or set of processes, to ensure normal body function.

TEMPERATURE CONTROL

Homothermic animals maintain a constant body temperature by a complex neuroendocrine homeostatic system. The resources used to conserve or lose heat vary considerably between species. Most mammals have a coating of hair to retain body heat, and also in some cases to prevent absorption of excessive external heat. Marine mammals without hair have substituted an insulating layer of subcutaneous fat. Sweating in many mammals is a physiological device for losing heat, and shivering a way to generate heat from muscle activity. All these are important in the overall maintenance of constant temperature.

The heat shock response is a particular defence mechanism against abnormally high temperatures. Temperatures of 42°–43° C induce the heat shock response in all mammalian cells that have been adequately studied. Specific proteins are synthesised in the induced response, and these produce profound changes in cell physiology. An animal that encounters excessive heat may well be in an environment where the temperature is liable to become even higher. The initial heat shock response is therefore an adaptation to improve the cells' defence against more extreme hyperthermia. Such protection has been demonstrated in many cases, although the mechanisms involved are not fully understood. It is possible that the heat shock proteins in some way protect other essential proteins from heat-induced damage. There are relationships between heat shock proteins and chaperonins, which are important in the control of polypeptide folding (see above, 'Removal of defective proteins'). So heat shock proteins may have a role in preventing heat-induced unfolding or denaturation of proteins. The response may also shut down metabolic and enzymic processes that are particularly heat sensitive. Another defence mechanism is the generation of heat in fevers, which is important in counteracting invading pathogens.

PHYSIOLOGICAL HOMEOSTASIS

Temperature control is, of course, a fundamental homeostatic mechanism that provides a uniform environment for innumerable metabolic processes. Homothermic animals are at an advantage over those that are poikilothermic. The latter are much less adaptable because they depend on the environment to provide the temperatures necessary for movement, feeding, reproduction and so on.

The study of the maintenance of physiological homeostasis is a discipline in its own right, and, as in the case of immunology, it would be inappropriate to attempt a review here. A few examples briefly illustrate the range of mechanisms of physiological controls that maintain the normal functions of tissue and organ systems. Many depend on signalling by hormones, and the control of glucose levels by insulin is a fundamental control mechanism. Neuroendocrine controls are also vital for many normal bodily functions. The sympathetic nervous system plays a central role in the normal functioning of the major internal organs, including food intake and utilisation. The response to muscular activity is an increased intake of oxygen and a greater heartbeat rate. The haematopoietic system is subject to many homeostatic mechanisms: shortage of oxygen results in enhanced production of blood erythrocytes; similarly, loss of blood from wounds stimulates the haematopoietic systems to make up the shortfall. Control of cell proliferation in many contexts depends on the activity of many specific growth factors, which are at present the subject of intensive studies.

The word 'homeostasis' is classically used in the context of stabilising physiological mechanisms, but it can also be used in the wider context of cellular and molecular mechanisms. If this is so, then most of the maintenance mechanisms discussed in this chapter have a homeostatic role in preserving cell integrity and functions (see also Chapter 9). It is commonly asserted that ageing is the result of a breakdown or failure of homeostatic mechanisms. This may well be the case, but homeostasis is probably too broad a term, and it is perhaps better to think of ageing as an eventual failure of maintenance *in toto* to preserve the normal functions of a body, the evolved design of which is incompatible with indefinite survival.

ORDER AND DISORDER

The unique property of living organisms is their ability to defy the laws of thermodynamics (Schrodinger 1944, and see Note 3.7). They exist by preserving order, and to do this they extract energy from the environment. Schrodinger refers to organisms 'feeding on negative entropy' to counteract the thermodynamic trend to increasing entropy. Organisms in a steady state can preserve themselves indefinitely by this means, provided the environment is appropriate. Mammalian organisms, and all others that become senescent and die, preserve order for only a limited period of time, even in an optimum environment. This time-span, I believe, depends on the efficacy of the maintenance mechanisms discussed in this chapter: *they act to preserve order.* Ultimately, senescence signals the gradual loss of order, an increase in entropy, and death is the return to disorder.

4

Theories of ageing

In Chapter 3 we saw that changes during ageing can occur in many components of the body, yet it is common to refer to *the process* of ageing, as if there was some underlying single mechanism that produced all these deleterious phenotypes. This is also often the case when a theory of ageing is proposed: in the formulation and presentation of the theory, the author often presupposes that there is one major mechanism or process that needs to be understood or explained. Thus, the significance of any particular theory is often overstated by its originator or supporter. The situation is made worse by the multiplicity of theories that have been discussed in this century. Many can now be ignored because they are outdated and have little relevance to modern knowledge. Medvedev (1990) has recently reviewed a large number of molecular and cellular theories, and has classified them into groups of related or overlapping proposals. It is probable that many of these have some basis of truth, and that none on its own 'explains' ageing. In this chapter some of the major proposals are reviewed, as are any pertinent experimental data that either confirm or are contrary to the theory's predictions. It is significant that the theories all relate in one way or another to the various maintenance mechanisms that were reviewed in the previous chapter.

SOMATIC MUTATIONS AND DNA DAMAGE

We know that DNA is a sensitive target in the cell, since there are only two copies of many essential genes. It is not surprising therefore that many authors have asserted, or even assumed, that defects in DNA are likely to be much more serious than defects in other cellular components, which may be present in multiple copies and can be readily replaced by metabolic processes. It is true that defects in DNA can be repaired, but repair is not infallible, and the failure of repair can lead to permanent impairment of gene function. Most somatic mutation theories of ageing do not in fact discuss the relationship between DNA damage per se and the fixation of this as a heritable change in DNA sequence (see Note 4.1). This is so because most of the theories were proposed before much information about accurate and error-prone repair was available.

Early ideas, such as those of Szilard (1959) and Curtis (1966), were concerned primarily with damage to whole chromosomes. Curtis spent many years testing the prediction that chromosome damage increased during ageing, and obtained evidence that this was the case in his studies of chromosomes in the dividing cells of regenerating liver. His results are perhaps not too surprising, but the problem is to assess the overall phenotypic significance of the changes seen. Although double-strand breaks in DNA can be repaired, we would expect some proportion to escape repair, and these would then be seen as chromosome breaks. (Note that attempted replication of a single-strand gap can produce a double-strand break.) Over a period of time the number of breaks or other chromosome abnormalities such as rearrangements would be expected to increase. This was confirmed in the detailed and comprehensive studies by Court Brown and colleagues (Jacobs et al. 1963; Jacobs, Brunton & Court Brown 1964; Court Brown et al. 1966) of the chromosomes of PHA (phytohaemagglutinin)-stimulated lymphocytes from human donors. These results demonstrate beyond doubt that chromosome abnormalities in lymphocytes increase during ageing.

Szilard was familiar with the target theory of genetic damage, namely that genes and chromosomes are susceptible to inactivating 'hits' from radiation, or from other damaging agents. His theoretical treatment of the somatic mutation theory is by far the most detailed that has been published. He made the assumption that many single-copy genes are linearly arranged along the chromosome, and that there is a pair of homologous chromosomes. He assumed mutations are recessive, so the inactivation of one gene would leave its partner intact, and in general this should have no or very little phenotypic effect. A serious cellular defect would be seen only when *both* copies of important genes were inactivated. Szilard's calculations show that this would be too rare an event to be an important cause of ageing. Instead, he concluded that the whole chromosome might be the sensitive target. There are only 23 chromosome pairs in humans, so the chromosome target is immeasurably greater than the target provided by individual pairs of genes.

Many studies have shown that whole-body irradiation significantly shortens the lifespan of mice (reviewed by Neary 1960; Comfort 1979). These results have been questioned on the grounds that irradiated animals may not develop the 'normal' features expected of premature ageing. In answer to this criticism, Lindop and Rotblat (1961) carried out a comprehensive study on the life-shortening effects of radiation, and the causes of death and other pathologies were investigated by the post-mortem examination of all control and irradiated animals. The pathological conditions in the irradiated animals were very similar to those in the controls, and they therefore concluded that the radiation did indeed induce premature ageing. These results therefore provided support for Szilard's or other somatic mutation theories of ageing.

Szilard's theory makes a strong prediction that the longevity in animals would be dramatically altered by a change in ploidy, that is, a change in the number of sets of homologous chromosomes. Haploid animals would be expected to have a very short lifespan, since the inactivation of one chromosome would be a lethal event, whereas triploid or tetraploid animals would have increased longevity in comparison to diploids, since three or four events would be needed to inactivate all homologous chromosomes. In response to Szilard's proposals, Maynard Smith (1959, 1962) pointed out that the relevant experiments had already been done with the wasp *Habrobracon,* where males are haploid and females are diploid. It was clear that the two sexes did not have very different longevities. Moreover, radiation treatment would be expected to have a very severe effect on the lifespan of males, and a much lesser effect on females. In fact, radiation-induced life shortening in males was much less than predicted from Szilard's theory, but it must be borne in mind that insects have few dividing cells in the adult, and so are very radiation resistant.

Somatic mutations have also been discussed in relation to the *in vitro* ageing of human fibroblasts (Holliday & Kirkwood 1981, and see Chapter 5). Again, it can be shown that recessive mutations on the autosomes, and also on the X chromosome (which is effectively haploid in females, as well as in males), could not account for *in vitro* ageing unless the mutation rate was extremely high. To explain the dying out of whole populations, the average viability of individual cells would also be expected to be very much lower than is actually seen during their growth. In addition, it is possible to obtain tetraploid fibroblasts and compare their lifespan with diploids (Hoehn et al. 1975; Thompson & Holliday 1978). These experiments clearly show that the two types of cell have the same or similar lifespans, which is contrary to the prediction of the somatic mutation theory.

Burnet (1974), in his book *Intrinsic Mutagenesis,* discussed the significance of somatic mutation in a variety of contexts, including age-related pathologies. He was well aware of the significance of DNA repair, and drew attention to the abnormal features of individuals known to have defects in repair mechanisms. In several repair-deficient syndromes, it is hard to know whether such individuals age prematurely, because they often die young from cancer or other causes. (Some of the relationships between inherited genetic defects, disease and ageing are further discussed in Chapter 8.) Other authors have questioned Szilard's assumption that almost all mutations would be recessive. Of all inherited diseases known in humans, of which there are a great number, a considerable proportion are known to be dominant, although perhaps with variable penetrance (see Note 4.2). If germ line mutations are often dominant, why should not the same be true of somatic mutations? Morley (1982) suggested that mutations might often be codominant, so that the heterozygote would have a significantly different phenotype from that of the normal homozygote. Thus, the gradual accumulation of such mutations during ageing would be expected to have increasingly severe effects. Another variation on this theme was Fulder's (1978) suggestion of a 'mutation catastrophe'. He argued that individual mutations might have no or very little effect, but as the number of mutations increased there would be a greater and greater chance of abnormal interactions, the loss of homeostasis and other age-related changes. The mutations could be occurring linearly with time, but the major phenotypic changes would not be seen until late in life. Such 'multiple-hit' models do explain in formal terms age-related diseases, as is well known, for example, in the case of human carcinomas (Cairns 1978). Burch (1968) took the extreme view of trying to explain the origin of all age-related diseases in terms of multiple-hit models as well as some, such as dental caries, that are not normally regarded as age-related. For any one disease, he could produce a formula that explained the incidence in terms of successive events, which could be mutations. (Burch's calculations provoked the wry comment that it would be possible to use exactly the same argument to explain the age-related election of British citizens to the Privy Council [but see Note 4.3].)

As well as chromosome abnormalities, it is now possible to study gene mutations in stimulated T lymphocytes. The gene coding for hypoxanthine phosphoribosyl transferase (HPRT) is X linked, and mutations that inactivate the single copy of the gene produce a cell that is resistant to 6-thioguanine (Morley, Cox & Holliday 1982). This method was exploited in the first quantitative studies of somatic mutations during ageing, and it was demonstrated that the mutations increased in T lymphocytes with age (see Fig. 4.1). The results taken together suggest that the rate of increase was exponential rather than linear, although the deviation from a linear increase was not statistically significant. An exponential or quasi-exponential increase in mutation frequency is of particular interest in the general context of error theories (see section on 'Protein errors' below), since it implies that mutations could result as a secondary consequence of other cellular defects during ageing. The inactivation of HLA-A alleles has also been studied in T lymphocytes from donors of different ages. The results show that there is an increase with age of mutations in this gene as well (Grist et al. 1992).

Some molecular biologists believe that the powerful methods of analysis that are now available can be used to study changes in DNA during ageing. By using appropriate probes, it is thought that changes in gene structure could be seen during ageing. For example, deletion or rearrangement of DNA sequences would be seen as an altered pattern of bands in a Southern blot. Unfortunately, this expectation is extremely unlikely to be borne out for the following reasons. Let us assume that mutations and changes in DNA do occur at a significant rate during ageing. The forward mutation rate of a particular gene is known to be low, perhaps 10^{-5}–10^{-6} per cell generation. Suppose this frequency is increased 100-fold during ageing, and this particular gene is then examined using Southern blots. Only one in 10^3 or 10^4 cells will have the mutation, whereas at least 99.9% of cells will have the normal gene. Clearly the abnormal molecules will not be detected. One could also imagine amplifying a specific region of DNA from single cells and then sequencing

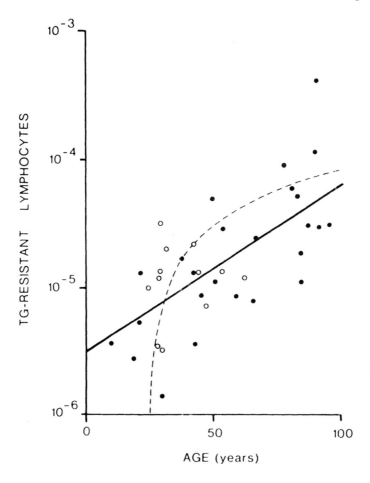

Figure 4.1. The increase in somatic mutation in human T lymphocytes with age. The phenotype scored is resistance to 6-thioguanine, and these cells lack the enzyme hypoxanthine phosphoribosyl transferase (HPRT): ●, non-laboratory donors; ○, laboratory donors. The regression line of best fit for logarithmically transformed data (—) corresponds to about 3% increase per year. The untransformed linear regression (- - -) gives a less good fit to the data. (Reproduced with permission from Morley et al. 1983. See also Trainor et al. 1984.)

the product. To detect a single deletion or rearrangement, one might have to sequence at least 1000 stretches of DNA. Point mutations might be somewhat more common, but an enormous amount of work would still be required. The project is frustrating, because every cell may actually contain many mutations, but because their distribution is random, they cannot be identified. A DNA change that can be measured is loss of telomeric DNA with age. This occurs during the ageing of cells in culture (see Chapter 5) and also *in vivo* (Lindsay et al. 1991).

An interesting variant of the somatic mutation theory is the idea that transposable elements might increase in number during ageing (Murray 1990). It is known that during transposition the parent copy remains in the chromosome, whilst a replicated copy can be inserted elsewhere in the genome. Thus, one copy gives rise to two, and two to four, and so on. Hence, in principle, the number of copies might increase exponentially with time, and some of these would inactivate important genes. It does seem true that specific DNA elements can increase rapidly in number during evolution of germ line DNA, such as Alu sequences in primate DNA. So it is possible that increases could occur also in somatic cells, even though the time scale is enormously shorter. The theory is at least testable, since a suitable probe could be used to measure copy number during ageing.

It was previously mentioned that some discussions of the somatic mutation theory of ageing do not distinguish between DNA damage and mutation. Normally, damaged bases are excised without mutation, but Lindahl (1993) has recently suggested that there could be important exceptions to this. He argues that some types of DNA base damage, which are relatively uncommon, are not a substrate for any repair enzyme. Such lesions might therefore accumulate in long-lived, non-dividing cells, such as neurones. In dividing cells they would simply be diluted out, and it could be assumed that they do not bring about mispairing of bases. (Any lesion that does induce mispairing is, of course, likely to be a substrate for a repair enzyme.) There is extensive discussion of DNA damage and ageing in the recent book *Aging, Sex and DNA Repair* by Bernstein and Bernstein (1991). Curiously, the authors concentrate on DNA damage that might affect transcription or DNA replication, and they are somewhat dismissive of the importance of mutations (i.e. DNA base changes) in ageing. They review a considerable body of evidence which indicates that DNA damage does accumulate in brain, muscle or liver with increasing age. Most of this evidence relates to DNA breakage, which can be measured with quite sensitive methods. Their main postulate is that DNA damage accumulates because repair is inadequate, especially in post-mitotic cells active in metabolism, such as neurones, but also in other types of cell. They propose that the inadequacy of repair leads to ageing, and the major function of meiosis, sex and outbreeding is to keep the germ line free of damage (see Chapter 6).

The experimental studies that have so far been carried out generally support the view that DNA damage, mutations and chromosome abnormalities increase during ageing (see also Rattan 1989). Moreover, some premature ageing syndromes in humans have an elevated incidence of genetic defects (see Chapter 8). It would be surprising if all these changes at the DNA level did not have some adverse effects and contribute to the phenotype of ageing mammals. Nevertheless, as we shall see, there are quite strong grounds for the belief that the accumulation of genetic defects may be just one of many causes of ageing.

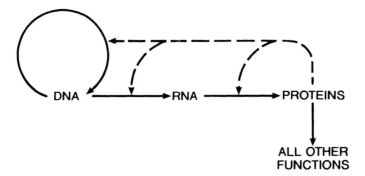

Figure 4.2. Error feedback in information transfer. The fidelity of DNA, RNA and protein synthesis depends on the normal function of many proteins. Since some errors occur in the synthesis of these proteins, there is a given probability of error feedback.

PROTEIN ERRORS

The protein error theory has perhaps provoked more discussion and experiment than any other. Medvedev (1962) first pointed out the likely significance of errors in macromolecules during ageing (see also Medvedev 1980), but the proposal by Orgel (1963) highlighted a particular problem that has to be solved both in the early evolution of organisms and in extant cells. The problem relates to the fidelity of information transfer from DNA to protein. If an enzyme is made with an error, that is, a single incorrect amino acid substituted for the normal one, then several consequences are possible: (1) the molecule may be unaffected or only slightly affected in function; (2) it may be completely inactive, or (3) it may have an important altered property, such as loss or change in substrate specificity. This third consequence will not matter for the majority of housekeeping enzymes, because the affected molecule will be only one amongst many normal ones, and eventually it will be degraded. The cellular problem relates only to that subset of enzymes or proteins that are actually involved in the transfer of information from DNA to protein. These include RNA polymerases, aminoacyl tRNA synthetases, ribosomal proteins and other components of translation. An error that changes the specificity of one of these molecules means that in carrying out its normal function it may introduce new errors, which would otherwise not have occurred. Imagine, for example, an erroneous RNA polymerase molecule, which increases the error frequency in transcription 10-fold: during the course of its lifetime in the cell, this molecule may be responsible for the synthesis of many mRNAs, each of which will code for protein with ten times the normal error frequency. The concept of error feedback in protein synthesis is illustrated in Figure 4.2. Orgel pointed out that error feedback could in principle destabilise the cell, so that as protein synthesis proceeds the error level will gradually rise, but at an exponential rate, so that eventually an 'error catastrophe' in protein synthesis will occur. Once protein error propagation has begun to occur, there is no cellular mechanism for reversing the process, and finally the cell will not be able to synthesise normal proteins. Orgel did not propose that this could be *the cause* of ageing, but instead that it might contribute to ageing. This theory is also of more general interest, because it shows that inheritance is not based only on DNA, since a cell must also inherit from its parent a normal machinery for the transfer of information from DNA to RNA and protein. If that machinery is defective, even completely normal DNA cannot function. In complete contrast to the somatic mutation theory, the protein error theory proposes that deleterious changes are based in the cytoplasm. Unfortunately, some authors have not understood the different origins and effects of errors in DNA and proteins.

Although RNA polymerase may be a sensitive target, errors in other RNA molecules such as tRNAs and ribosomal RNAs may also be important. These could occur in the primary sequence, or in the many post-synthetic modifications to bases in RNA molecules that are known to occur. In addition, one must take into account errors in splicing of RNA transcripts, which in some cases would lead to the insertion or deletion of amino acids. It should be noted that the accuracy of transcription and translation is about 10^4 less than the accuracy of DNA replication. If the error level in protein synthesis is as high as 10^{-3}, as some estimates suggest, then every molecule of a large protein with a thousand amino acids will contain, on average, one error. Detailed studies in *E. coli* suggest that every ribosome contains several molecular errors (Kurland 1987).

There have been many theoretical discussions of the likelihood of error feedback in protein synthesis, which are briefly reviewed in Note 4.4. In summary, one can state with some certainty that two possibilities exist. Either the cell is in a steady state, where there is a given level of spontaneous errors and a given unknown level of error feedback that is insufficient to bring about error propagation; or the cell is in a metastable or unstable state, where there is a given possibility that error feedback will begin to escalate (Fig. 4.3). Once this has occurred, the cell will eventually become non-viable. It was emphasised in Chapter 3 that energy and resources are required to synthesise macromolecules with accuracy. There must therefore be some optimum level of accuracy in RNA and protein synthesis. The evolution of this optimum depends on the need to (1) make a large proportion of normal functional molecules; (2) avoid error propagation in protein synthesis, and (3) minimise the energy requirements for proof-reading and accuracy in the synthesis of RNA and proteins. The crucial question, therefore, is the 'distance' the optimum has to be from the likelihood of initiating an eventual error catastrophe (for further discussion, see Note 4.4). The likelihood of error propagation is also important in the early evolution of life. Initially, the mecha-

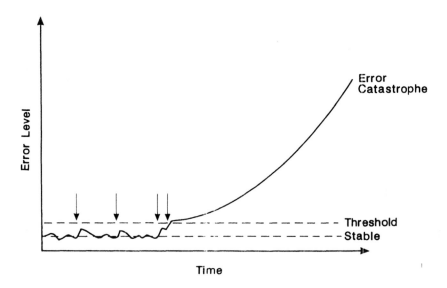

Figure 4.3. Under normal circumstances the level of errors is below a threshold value, so a steady state level exists, with random perturbations. There is a given probability that critical errors might occur (for example, in RNA polymerase), which would destabilise the machinery for protein synthesis, leading to an irreversible increase in the error level.

nisms of information transfer must have been very inaccurate, so how could error feedback be avoided, or to put it another way, how did an inaccurate system generate an accurate one? This problem of the evolution of life and the problem of the maintenance of a functional complex organism are not totally unrelated, but obviously they cannot be pursued further here.

In present-day organisms, it is important to know whether an error catastrophe can be induced artificially. The strongest evidence that this is the case comes from experiments with *Escherichia coli.* In two studies the aminoglycoside antibiotic streptomycin has been used to increase errors in the synthesis of proteins artificially (Branscomb & Galas 1975; Rosenberger 1982). Conditions could be found where the protein error level steadily increased whilst cells were outwardly normal, but finally there was a dramatic loss of cell viability. In both studies, β-galactosidase was used to monitor the accuracy of protein synthesis, and it could be shown that errors were increasing exponentially until the cells died. There is also evidence that an error catastrophe can occur in the temperature-sensitive *leu*-5 mutant of *Neurospora* at the restrictive temperature (Lewis & Holliday 1970).

The protein error theory of ageing in its modern form proposes that a cell is normally in a steady state, but there is a given probability that it will cross a threshold to an unsteady state, with increasing errors in proteins (reviewed by Kirkwood et al. 1984; Rosenberger 1991). There has been much misunderstanding and misrepresentation of the theory, which is in part due to the introduction of the term 'error catastrophe'. This has led some to believe that the theory predicts that all cells in ageing animals will have very high levels of abnormal proteins, which should be very easy to detect. In reality, a tissue may have an aged phenotype if some of its cells have a small proportion of abnormal proteins, which are hard to detect (see below).

A theory is only valuable if it can make testable predictions, and as has often been stated, the protein error theory does make several specific predictions. What has much less often been realised is that several of these predictions are very hard to test. One prediction is that treatments that induce errors in proteins may accelerate ageing. For example, as originally suggested by Orgel (1963), young animals fed for a while with an amino acid analogue in their diet might have a shorter lifespan than untreated animals. A second prediction is that senescent cells should contain abnormal protein molecules. A third is that the actual accuracy of synthesis of polypetides should be lower in old cells or organisms than in young ones. A fourth is that mutations that are known to reduce the accuracy of synthesis of proteins, by whatever means, should also accelerate ageing. Conceivably, mutations might increase the accuracy of synthesis, and have the reverse effect on ageing.

These predictions have been tested in various experimental organisms with various degrees of success. Amino acid analogues, 5-fluorouracil and aminoglycoside antibiotics (which interact with ribosomes and reduce the accuracy of translation) have been tested with fungi, *Drosophila*, mice and cultured human fibroblasts. These results are summarised in Table 4.1. Life-shortening effects have been seen, but these results are almost always open to more than one interpretation. For example the incorporation of amino acid analogues into the protein of *Drosophila* larvae, as well as their possible reutilisation in later protein synthesis, could lead to abnormal proteins with a much shorter survival time than normal long-lived molecules, or alternatively, to abnormal insoluble proteins. Both these consequences might reduce longevity. The results with the aminoglycoside antibiotic paromomycin are the most interesting. It induced premature senescence of human fibroblasts, and this was not reversed by returning the cells to normal medium. One criterion of senescence is the accumulation of autofluorescent material (Rattan et al. 1982). Cells treated with paromomycin had an elevated amount. In the same study, it was shown that Werner's syndrome cells, which have a short lifespan, have a high level of autofluorescence.

Early attempts to detect the accumulation of abnormal protein molecules in old organisms depended heavily on immunological methods. Antibodies are raised to a known enzyme, and the ratio of enzyme activity to the amount of cross-reacting material is measured. If non-functional molecules are present, some will react with the antibodies, so to inactivate a given amount of en-

Table 4.1. *The effects on ageing of treatments that might reduce the fidelity of protein synthesis, or otherwise alter proteins*

Organism	Treatment	Result	Reference
Podospora anserina	Ethionine or pFPA[a] in growth medium	Lifespan reduced	Holliday (1969)
Drosophila	Mixture of amino acid analogues in diet of larvae	Lifespan reduced	Harrison & Holliday (1967)
	Streptomycin[b] in diet of larvae	Lifespan reduced	Harrison & Holliday (1967)
	pFPA fed to adults	No effect	Dingley & Maynard Smith (1969)
Inbred mice (CBA)	pFPA in drinking water of young mice (4-week treatment)	Moderate doses reduced lifespan; higher doses did not	Holliday & Stevens (1978)
Human fibroblasts	Ethionine and pFPA in growth medium	No effect	Ryan, Duda & Cristofalo (1974)
	pFPA in growth medium	No effect	Tarrant & Holliday (unpub., but see Holliday 1991b)
	Fluorouracil in growth medium	Premature senescence; lifespan reduced	Holliday & Tarrant (1972); Holliday (1975)
	Paromomycin[c] in growth medium	Premature senescence; lifespan reduced	Holliday & Rattan (1984)

[a]pFPA, para-fluorophenylalanine.

[b]It was believed that streptomycin would affect the fidelity of translation by cytoplasmic ribosomes. In fact, it does not act on eukaryotic ribosomes and is more likely to affect mitochondrial protein synthesis.

[c]The related antibiotic paromomycin does reduce the fidelity of translation by eukaryotic ribosomes. Neither the fluorouracil nor the paromomycin treatment initially affected cell growth rate (see also Note 9.2).

zyme it is necessary to add more antisera. By this means the accumulation of abnormal protein was demonstrated in ageing nematodes, and also in the liver tissue of ageing mice (reviewed by Rothstein 1975, 1977, 1979, 1982). The interpretation of these results was that a particular inactive form of the enzyme was present, rather than a collection of different abnormal molecules. However, in no case has an abnormal protein been identified and characterised. The results, nevertheless, have been used to discredit the protein error theory. The argument seems to run as follows. First, the error theory predicts that abnormal proteins will accumulate during ageing; second, abnormal proteins have been detected, but they do not arise from random errors. Therefore the error theory is incorrect. The possibility remains that some proteins are modified and inactivated during ageing (see the section 'Protein modifications', below), but there is nevertheless *also* an accumulation of random errors.

Another indirect method of detecting abnormal molecules depends on the fact that many amino acid substitutions do not affect enzyme activity, but do reduce the thermal stability of the molecule. Thus, the heat inactivation of a homogeneous collection of protein molecules gives an exponential decay in activity, but the presence of a fraction of heat-labile molecules is demonstrated by a rapid initial loss of activity, followed by an exponential decay. The enzymes glucose-6-phosphate dehydrogenase, triosephosphate isomerase and 6-phosphogluconate dehydrogenase have been used in various studies, and the existence of a heat-labile fraction has been shown in ageing cells or tissues (Holliday & Tarrant 1972; Holliday, Porterfield & Gibbs 1974; Tollefsbol, Zaun & Gracy 1982; Holliday & Thompson 1983). Again, there are at least two interpretations of these results. Both random errors and the post-synthetic modification of the enzyme to an active but less stable form of the enzyme would give the observed results.

These indirect tests are inadequate, and what is needed is a direct measurement of errors in protein synthesis. The first requirement is to measure the accuracy of protein synthesis in normal cells, and the second is to monitor any changes in accuracy during ageing. This poses considerable problems for the experimentalists, and to this day no direct test of the protein error theory has been carried out. The difficulty is similar to the one that was raised concerning the detection of random mutations in DNA by molecular methods. Protein errors will be randomised, so there might, for instance, be 90% normal molecules and 10% with errors; but this 10% will consist of a heterogeneous collection of erroneous molecules. Standard methods of analysis such as electrophoresis, isoelectric focusing, amino acid analysis, chromatography or peptide mapping will not detect these levels of abnormal molecules. Sequencing the protein illustrates the problem. Suppose every molecule has at least one error, but these are scattered along the polypeptide chain; then normal procedures will produce the wild-type sequence, because it is the consensus, and no errors will be seen.

The technique that has been attempted most often is to look for the mis-incorporation of a radiolabelled amino acid in a peptide or protein in which it is not normally present (reviewed by Kirkwood et al. 1984). Thus, the errors are detected as counts in purified peptide or protein. This procedure was pioneered by Loftfield (1963) using peptides from ovalbumen, and later from globin (Loftfield & Vanderjagt 1972), and it has been successfully applied to bacterial systems (Edelmann & Gallant 1977a). A major problem of the technique is the necessity of achieving almost complete purity of the protein or peptide, with the exclusion of all contaminating material, since this would contain the labelled amino acid. Suppose errors do occur and 1% of molecules contain, on average, one incorrect labelled amino acid. (Note that this 1% represents the errors attributable to misincorporation of only *one* amino acid; for all amino acids the error level would be roughly 20-fold higher.) Purification of a protein to 99% leaves 1% contaminating protein material, and each of these molecules is likely on average to contain several residues of the labelled amino acid. This would swamp the 1% of erroneous molecules the experimentalist is trying to detect. To detect the errors, the protein must be purified to at least 99.9%, which is a far greater degree of purity than most protein chemists would hope to achieve. An interesting experimental approach involves using purified mRNA for a virus coat protein that lacks codons for cysteine and methionine (Luce & Bunn 1989). Extracts of human diploid fibroblasts are used for *in vitro* translation, and the product is purified by immunoprecipitation. The error frequency was over six-fold higher using extracts from senescent cells than with extracts from young cells. An important control showed that the antibiotic paromomycin, which reduces the accuracy of translation, induced errors in young-cell extracts.

Because direct measurements are so difficult, indirect procedures have been used to attempt to direct misincorporation. One depends on the change in charge that is seen in 'stuttering' of proteins on a two-dimensional (2-D) gel (see Note 4.5). It is known that when a high level of errors is induced artificially (for example, by using aminoglycoside antibiotics that induce mistranslation), then the major protein spot on a 2-D gel is accompanied in one dimension by minor spots, with an increase or decrease of one charge, and fainter spots with an increase or decrease of two charges, and so on (see Parker et al. 1978). Only the misincorporation of a charged amino acid for an uncharged one, or vice versa, produces a molecule with altered mobility. From experiments that have been done with *Drosophila* and human cells, it is clear that 2-D protein gels prepared from ageing cells or tissue do not demonstrate stuttering, which shows that the error level is not enormously high. This method was used in a paper with the title 'Protein synthetic errors do not increase during the aging of cultured human fibroblasts' (Harley et al. 1980). In fact, the study documents only the production of stutter spots under abnormal physiological conditions induced by histidine starvation, and then claims to deduce the 'real' error level in young and old cells. There is no actual measurement of these real values (see Note 4.6). Another indirect method depends on the examination of the properties of viruses grown in senescent cells. In one important study, no effects on viruses were seen (Holland, Kohne & Doyle 1973). As was previously mentioned, some authors have taken the term 'error catastrophe' too literally, and have concluded that the theory predicts high levels of an erroneous protein in ageing cells. Unfortunately, we are woefully ignorant of the effects of altered proteins on cellular physiology and function, either at the level of the single cell or in tissues made up of populations of cells. Would 1% or 10% of erroneous protein molecules have serious effects on cellular function? Would 1% or 10% of cells containing significant levels of altered proteins have serious effects on tissue function? We do not have answers to these questions.

In conclusion, it can be stated that:

1. an error catastrophe with a very large proportion of altered molecules is not seen in the proteins of ageing cells;
2. the major prediction of the protein error theory has not yet been tested, owing to experimental difficulties;
3. the evidence for the theory so far obtained has been with indirect methods and the results are open to more than one interpretation, and
4. there is no information about the likely effects of protein errors on normal cell and tissue function.

PROTEIN MODIFICATIONS

Classical studies of ribonuclease demonstrated that the complete denaturation of a protein to a polypeptide chain could be followed under different conditions by its spontaneous reassembly to an active enzyme. In other words, the sequence of amino acids determines the way a protein will fold up to produce its normal three-dimensional structure. The discovery of altered inactive enzymes in old animals led Rothstein (1975, 1979) to question the dogma of protein self-folding. He suggested that active enzymes are often folded into a high energy state, and that they have a given tendency to revert to a lower energy form that is inactive, but retain epitopes that react with specific antibodies. Normally such inactive molecules would be degraded by proteases, but if protein metabolism is abnormal in ageing cells, the altered proteins would accumulate (Makrides 1983; Hipkiss 1989). More recently, a class of proteins has been discovered called chaperonins (Ellis & van der Vies 1991; Gething & Sambrook 1992). These have a role in bringing about the correct folding of other proteins, and also in transporting them to appropriate locations in the cell. The existence of chaperonins shows that the work with ribonuclease is an oversimplification, in that not all proteins fold up spontaneously, and also makes Rothstein's suggestion much more plausible.

If the theory is broadened to include various forms of protein modification, then it is supported by a considerable body of evidence, which can only be briefly reviewed here. Possible protein modifications are listed in Table 4.2. Early experiments showed that inactive enzyme cross-reacting material accumulated during the ageing of the nematode *Turbatrix aceti*. The enzymes affected were isocitrate lyase, phosphoglycerate kinase, enolase and aldolase. Later, similar results were reported using mouse tissues, but not all enzymes examined were altered. The altered enzymes were not changed in molecular weight, charge, terminal amino acid residues or –SH groups, but specific ac-

Table 4.2. *Possible alterations to proteins during ageing*

1. Primary errors in transcription and translation
2. Deamidation of asparaginyl and glutaminyl residues
3. Non-enzymic glycosylation; advanced glycation end-products (AGEs)
4. Oxidation
5. Cross-linking of lysine residues
6. Proteolytic cleavage, without further breakdown
7. Partial denaturation (to a lower energy level)
8. Racemisation of L-amino acids
9. Abnormal phosphorylation
10. Abnormal methylation
11. ADP ribosylation

Note: See text and, for a recent review, Rattan, Derventzi & Clark (1992).

tivity was reduced, with corresponding appearance of inactive cross-reacting material, and the enzyme might have increased heat lability and sensitivity to proteases (Rothstein 1982). These changes were interpreted to mean that the enzyme molecules, at least in some cases, had undergone a conformational change, but other post-synthetic alterations were also possible. Gracy and colleagues (Yuan, Talent & Gracy 1981; Tollefsbol et al. 1982) have obtained evidence that deamidated forms of triosephosphate isomerase are less stable than the normal enzyme, and accumulate in aged cells and the eye lens. They suggest that this may be due to a failure of normal proteolysis during ageing.

As well as deamidation, proteins are subject to non-enzymic glycosylation (Cerami 1986; Lee & Cerami 1990; van Boekel 1991). The initial reaction involves amino groups, particularly that in the lysine side chain of proteins, and the aldehyde group of sugars to form a ketoamine or Amadori product (Note 4.7). This is subject to a series of higher-order reactions (collectively known as the 'Maillard reaction'), the chemistry of which is not understood, to give a complex mixture of products, which are referred to as advanced glycation end-products (AGEs). Long-lived proteins such as collagen and crystallin can become glycated, and the thickening of basement membranes seen in diabetics is also in part due to glycation. Human collagen with increased glycosylation forms fluorescent derivatives, and the amount of this material increases with age (Monnier, Kohn & Cerami 1984). Glycated proteins are likely to form cross-links with other proteins, and the upshot is the formation of material that cannot be degraded and will accumulate in cells. Protein oxidation is an important change listed in Table 4.2, and it is discussed in the next section. A number of modified forms of histones have been demonstrated during the ageing of rat tissues (Medvedev & Medvedeva 1991). The normal phosphorylation of proteins is essential for cell regulation mechanisms and for signal transduction. It is possible that abnormal phosphorylation would have serious effects on cells, but direct information is lacking.

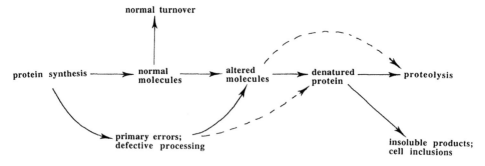

Figure 4.4. The protein modification theory of ageing. Altered enzymes may retain activity, but lose stability. Alternatively they may partly unfold to produce inactive molecules that react to antibody to the enzyme. Many types of protein modification are possible (see Table 4.2). All these molecules are subject to proteolysis, as are fully denatured molecules. In normal situations there is a balance between the formation and removal of altered proteins, but during ageing there could be an increase in their formation without increased proteolysis. Alternatively, proteolysis may be reduced. In both cases, the balanced or steady state would be impaired and altered molecules would accumulate. (Based on Rothstein 1975.)

The protein modification theory is summarised in Figure 4.4. In a normal steady state situation, the formation of altered protein molecules will be balanced by their proteolytic degradation. A problem arises if the formation of abnormal proteins is faster than their removal, or alternatively, if the normal proteolytic mechanisms are impaired. This is compounded by the tendency for abnormal molecules to aggregate, cross-link, and form insoluble, non-degradable inclusion bodies in cells. There is much evidence that long-lived, post-mitotic cells in various tissues are affected in this way, and this is probably an important component of the overall ageing process.

OXYGEN FREE RADICALS AND MITOCHONDRIA

Harman (1956) first proposed that short-lived oxygen free radicals might be an important cause of ageing (for a recent review, see Harman 1992). Ionising radiation produces tracks of free radicals in cells, and research after the Second World War demonstrated that radiation reduced lifespan (see 'Somatic mutations and DNA damage', above). However, early attempts to increase the lifespan of rodents with antioxidants were not successful (see Schneider & Reed 1985). Interest in the theory declined, but there is now a resurgence of research activity, and many favour the possibility that oxygen free radicals are an important cause of disease and ageing (see Halliwell 1987; Halliwell & Gutteridge 1989; Ames & Gold 1991; Ames & Shigenaga 1992; Ames et al. 1993).

It is becoming clear that the oxygen free radical theory stated in its broadest form overlaps with, or is part of, other theories of ageing. The spontaneous mutation or DNA damage theories of ageing do not necessarily invoke specific causes of damage, but the free radical theory proposes that DNA is a specific target for free radical attack. If this is so, then 'spontaneous' damage may in large part be due to oxygen free radicals. Similarly, the protein modification theory proposes that post-synthetic abnormalities gradually accumulate during ageing, but one of the major types of abnormality is oxidation by free radicals. Also, lipid peroxidation not only gives rise to insoluble and nondigestible age pigments such as lipofuscin, but also generates mutagenic lipid epoxides.

It has also been proposed that chronic infections can lead to degenerative disease, mediated by the release of damaging free radicals. Macrophages and other cells respond to invading bacteria, viruses or parasites, by releasing toxic or cytostatic free radicals (O^-_2, H_2O_2, NO and OCl$^-$). The resulting severe or chronic inflammation brought about by these host–pathogen interactions may lead to cancer and other degenerative diseases (Ames et al. 1993). It is widely believed that the release of oxygen free radicals, either as a by-product of normal metabolism or associated with inflammatory reactions, can contribute to a number of human age-related diseases, such as cardiovascular disease, decline of the immune function, rheumatoid arthritis, brain damage and cataracts (see Halliwell 1987; Halliwell & Gutteridge 1989; Ames et al. 1993).

It is thought that protein oxidation by free radicals is a major factor in these diseases. Such oxidation often occurs at specific metal-binding sites in the protein. Normally such a site would be occupied by magnesium or zinc, which is important for the function of the protein. However, iron can also bind to such a site, and this catalyses the Fenton reaction with hydrogen peroxide to produce the highly reactive hydroxyl radical (see Fig. 3.1). This can induce both structural and functional changes in proteins. An important product of the reaction is the carbonyl group, which is not normally present in proteins. There have been several studies of protein oxidation in relation to ageing, and there is evidence that the carbonyl product of oxidation increases in the ageing brain, eye lens and rat hepatocytes (reviewed by Oliver et al. 1987; Stadtman 1992). Such altered molecules could be included in the pathways shown in Figure 4.4, but it is also possible that they damage other molecules. Dean and colleagues (Dean et al. 1992; Dean, Gieseg & Davies 1993) have suggested that protein hydroperoxides or other reactive forms may diffuse to other parts of the cell and thereby produce progressive changes that may be important in ageing. This provides a positive mechanism for inducing damage, whereas oxidised proteins per se are rather unreactive. As was mentioned previously, the protein oxidation theory of ageing overlaps with the discussion of modified proteins in the previous section. A crucial question is the relationship between the rate of oxidation and the rate of removal of the abnormal molecules by proteases. The accumulation of oxidised proteins during ageing may be due to a reduction in protease activity.

Oxygen free radicals can also produce lipid peroxidation, and when this occurs in organelle membranes the outcome may be the generation of lipofuscins. These pigments are well known as biomarkers for ageing, since they accumulate in a wide variety of locations, including the brain, retina, heart, skeletal muscle and liver. Lipofuscin is probably a complex mixture of abnormal protein and lipid that has defied chemical analysis, and its effects on cell function are far from clear (see also Chapter 5).

Much attention has been paid to the effects of oxygen free radicals on DNA. It is clear that a variety of abnormal base adducts can be formed (see Chapter 3 and Note 3.4), and these are removed by repair enzymes with execution of the free bases or nucleoside in urine. A relationship between metabolic rate and the amount of thymine glycol excreted has been demonstrated in four species (Adelman et al. 1988), suggesting that the consumption of oxygen per kilogram of body weight determines the amount of DNA damage (but see Chapter 7). Recently, Lindahl (1993) has suggested that the estimates of free radical damage in DNA may be two or three orders of magnitude too high.

What is not in doubt is the greater damage in mitochondrial DNA, in comparison to nuclear DNA (Richter et al. 1988). Nuclei have very little oxygen metabolism, and it is a reasonable supposition that respiration is carried out by a specific organelle, the mitochondrion, to protect nuclear DNA from free radical damage. The mutation frequency of mitochondrial DNA is at least 10 times higher than in chromosomal DNA, which provides evidence of the importance of free radical attack, although inadequate repair could also be responsible.

Not surprisingly, it has been suggested that mitochondrial damage may be an important cause of ageing (Linnane et al. 1989). There have now been many studies of mitochondria during ageing (see Joenji 1992), and there is increasing evidence that deletions can arise in mtDNA. This work is based on the polymerase chain reaction (PCR) amplification of DNA, which can detect a proportion of molecules with a specific deletion, but cannot detect random mutations (see 'Somatic mutations and DNA damage', above). However, the relationships among mitochondrial turnover, loss of respiratory function and ageing are far from clear. One interesting possibility is that mitochondria are subject to a feedback loop, in which initial damage is continuously increased (Fig. 4.5).

Attempts to test the free radical theory have not been very successful. Early experiments by Harman suggested that antioxidant treatment could increase the lifespan of mice, but this has not been supported by subsequent studies (see Schneider & Reed 1985). Indeed, in view of the current interest, the absence of reports of life-extending effects of antioxidants suggest that such experiments have given negative results and have not been published. There

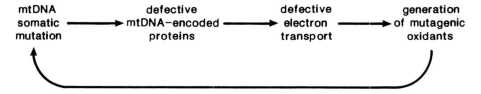

Figure 4.5. A feedback loop involving mitochondrial metabolism. Mitochondrial DNA has a high mutation frequency, which will produce altered proteins. This could increase the formation of oxygen free radicals, with further damage to DNA. (Reproduced with permission from Beckman, Hagen & Ames 1992.)

will always be strong pressure to publish experiments that demonstrate life extension.

The evidence for the validity of the theory rests on the demonstration that oxygen free radicals damage DNA, lipid and protein; that abnormal mitochondria increase during ageing; and that ionising radiation accelerates ageing. Also, it is suggestive that an additional gene for superoxide dismutase (Cu Zn SOD) in trisomy 21 (Down's syndrome) is associated with several features of premature senescence, including Alzheimer's disease (see Chapter 8). Excess SOD could result in an imbalance in the pathways of free radical removal (see Fig. 3.1). The specific activity of SOD is not greater in long-lived species, but the ratio of SOD activity to metabolic rate does correlate with species longevity (Tolmasoff, Ono & Cutler 1980). Since metabolic rate is related to the production of oxygen free radicals, SOD would be more effective in removing superoxide in long-lived species. Also, auto-oxidation andperoxide formation in brain and kidney tissues is inversely correlated with species lifespan (Cutler 1985). In another study,the concentration in serum of the antioxidant β-carotene was measured in several mammalian species. In this case, the concentration of β-carotene correlated with longevity (Cutler 1984).

All these observations provide support for the oxygen free radical theory of ageing; but as we saw in Chapter 3, there are effective defences against free radical damage, and if these defences are successful during the major part of an organism's life, why do they become inadequate during senescence? Does some of the damage inflicted simply accumulate with time, and significant phenotypic changes are only seen when the damage reaches a certain level? Or is it more likely that the ability of an organism to deal with free radical damage eventually declines? If the latter is correct, then the free radical damage during ageing becomes a secondary consequence of other primary causes of ageing.

IMMUNOLOGICAL THEORY

In his book *The Immunologic Theory of Ageing,* Walford (1969) argues that many pathological features of ageing can be attributed to dysfunction of the immune system, particularly auto-immune reactions. The book was written at a time when many of the basic features of the immune response were understood, but a great deal more has been discovered since that time, and it would be valuable to have an updated theory in the light of these developments (but see also Walford 1974). The efficiency of the immune system critically depends on the ability of the organism to distinguish self from non-self antigens. He writes:

> the immune theory regards ageing in vertebrates as an active autocatalytic process concerned with dividing cell populations leading to self destruction. . . . Ageing is due to somatic cell variation, particularly of those factors which determine self recognition. In higher animals the cells of the reticuloendothelial system are especially involved. Ageing in these species is brought about by the unleashing of self destroying processes of the nature of autoimmunity or transplantation disease. (pp. 198, 203–4)

Burnet (1974) has also argued that somatic mutation in cells of the immune system may have severe consequences. However, Walford does not refer only to somatic mutation, but to other possible sources of variation, and nowadays one could include epigenetic variation, perhaps based on changes in DNA methylation. The upshot of such variation is to release weak histocompatibility reactions, much less obvious than the direct rejection of foreign tissue. These reactions could have pathological effects on a variety of cells or tissues.

The decline in the immune response during ageing is well documented (reviewed in Nagel, Yanagihara & Adler 1986; also in Miller 1990). Many clinical observations have confirmed this decline. There is a progressive quantitative and qualitative loss in the ability to produce antibodies. There is a decline in T lymphocyte function with age. Indeed, the thymus, which processes the T cells, is often thought to be an organ strongly tied to the ageing process, and which can serve as a model for ageing overall. The human thymus reaches its largest size during adolescence, and then progressively atrophies. By age 50 it is about 15% of its maximum size. Many experimental studies have demonstrated the decline in function of lymphocytes during the ageing of mice. The evidence suggests that these changes are intrinsic to the cells themselves, rather than being imposed on the cells by extrinsic factors (Goodman & Makinodan 1975; Makinodan et al. 1976; Makinodan 1979; Makinodan & Kay 1980). A decline in the immune response is not the same thing as a failure to discriminate self from non-self. Walford's theory predicts that B lymphocytes produce antibodies to self antigens during ageing, or that cytotoxic T cells lose their ability to distinguish the presentation of foreign antigens by macrophages from that of self-antigens. Nevertheless, since the whole system depends on molecular specificity, it would be surprising if abnormal events did not occur in senescent animals. What is uncertain is their likely contribution to ageing.

There has been much discussion of the division potential of cells of the haematopoietic system *in vivo*. Stem cells appear to have sufficient division potential to last several mouse lifespans (Harrison 1979, 1984), but when such cells are passaged through several animals they eventually lose their ability to repopulate the bone marrow of a recipient animal. Also, it has been demonstrated that antibody-producing memory cells have a limited proliferative lifespan *in vivo*, estimated at about 100 divisions (Williamson & Askonas 1972, and see also Chapter 5).

The key feature of the immunological theory is that deleterious events, whether they be mutations, chromosome abnormalities or epigenetic changes, may have severe consequences for many other types of cells. The same type of genetic event in these target cells may have a much lesser effect, that is, some cell loss without necessarily much loss of function or influence on other tissue or organ systems. On the other hand, the immunological theory does not address the important issue of the origins of cell variability, which is presumed to result in the loss of recognition of self and non-self antigens. In this regard it is a secondary theory, based on other ideas about the generation of errors or defects in macromolecules.

DYSDIFFERENTIATION OR EPIGENETIC THEORY

Most cells in the adult organism are highly specialised and stably maintain their phenotype. A possible basis for this stability was briefly discussed in Chapter 3. Cutler (1982) proposed that one of the things that may go wrong during ageing is the 'dysdifferentiation' of cells. By that he meant that cells, instead of producing their normal complement of luxury proteins, sometimes synthesise proteins that are inappropriate. This ectopic expression of proteins might in many cases be harmless, but in other cases will be deleterious and perhaps further destabilise the cellular phenotype. It should be remembered that in the development of an organism, cells are assigned to specific developmental pathways. The result is a branching developmental tree, where cell type A produced in one tissue or organ is completely separated in time and space from cell type B, which is located elsewhere. Thus the specialised functions of A and B cells have evolved separately, and at no time do luxury proteins of A come into contact with those in B. An abnormal event that gives rise to an A type protein in cell type B, or vice versa, might produce an incompatibility, especially if the protein in question had an important regulatory role. The theory therefore proposes that there are regulatory defects that bring about ectopic expressions of proteins with specialised functions. Since we do not know how the stability of specialised cells is stably maintained, there is not a great deal that can at present be concluded about the importance of the theory.

One possibility is that DNA methylation may be involved (Holliday 1984a, 1987; Mays Hoopes 1989). We know that patterns of DNA methylation are very specific and stably inherited through the activity of a maintenance methylase. The level of total methylation declines during ageing. This is true for cultured cells *in vitro* (Wilson & Jones 1983; Fairweather, Fox & Margison 1987) and also *in vivo* (Wilson et al. 1987), although the extent of loss seen is barely measurable in the latter. The evidence does indicate, however, that DNA methylation in mice declines *in vivo* at a greater rate than DNA methylation in humans, which suggests that the rate of decline relates to longevity. It is improbable that all methylation has a role in gene regulation, but if we consider the subset of 5-methyl cytosine (5-mC) residues that are important, then it is possible that their loss could lead to ectopic expression of proteins. The evidence that specialised cells do produce ectopic proteins is minimal, but it is probable that it has never been sought systematically. With modern procedures for detecting antigens with fluorescent antibodies, or other sensitive methods, and the use of a flow cytometer, it should be possible to screen populations of differentiated cells for occasional cells that produce a protein normally seen only in some other specialised cell type. Such experiments could be done with appropriate cells from young and old animals.

The best evidence of epigenetic changes during ageing comes from studies of the X chromosome in female mice. It is known that the inactive X is normally stably maintained during somatic cell division (see Gartler & Riggs 1983). In translocations in which part of an autosome is attached to an inactive X, the inactivation can spread into the autosomal region (see Note 4.8). For example, in Cattanach's translocation there is an insertion of part of an autosome into the X chromosome. This autosomal region carries two genes that are important for coat colour, so if the region is inactivated the animal is albino, provided the normal autosomal genes are non-functional. Cattanach (1974) observed that as the animals aged they became more pigmented (the opposite of the usual ageing effect!), as shown in Figure 4.6a. This demonstrated that the epigenetic control of the inactivation was decreasing with age. More direct experiments were subsequently done using an inactive X-linked gene (coding for ornithine carbamoyl transferase), and it has been shown that this gene as well can become reactivated with age. The clearest result came from the study of individual cells, using a histological assay (Fig. 4.6b). A third example of age-related X chromosome reactivation has also been documented (Brown & Rastan 1988).

Housekeeping genes on the X chromosome have CpG islands (see Bird 1986), and several have been studied with respect to their DNA methylation. Whereas CpG islands are usually unmethylated, those on the inactive X are

Figure 4.6 (*facing*). Gene reactivation during ageing. (a) Female mice with Cattanach's and Searle's translocations, which are not mosaic for X chromosome activity. The albino animal (*left*) is 3 months old, but the same animal at 8 months has become pigmented (*right*). This is due to the activation of a tyrosinase gene inserted into the inactive X chromosome (see Note 4.8). (b) A histological assay that demonstrates the reactivation of a normal gene coding for ornithine carbamoyl transferase in female mice. The normal gene is on the X

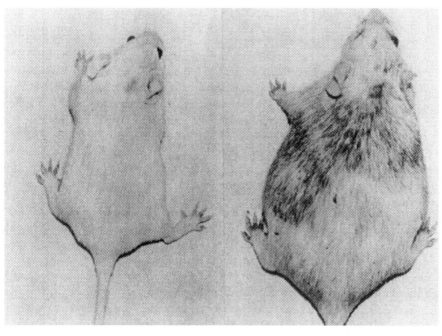

a

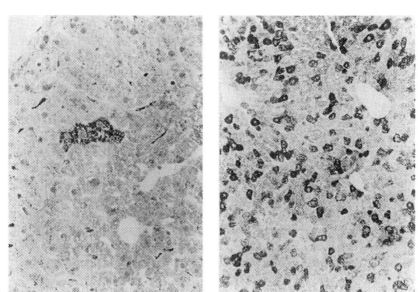

b

Caption to Figure 4.6 (*cont.*)

chromosome, which is always inactivated because the mice carry Searle's translocation and are therefore not mosaic, and the active X chromosome has a nonfunctional gene: (*left*) liver tissue of a young adult; (*right*) liver tissue of an old animal. (Part (a) reproduced with permission from Cattanach 1974. Part (b) reproduced with permission from *Nature,* Wareham et al. 1987; copyright © 1987 Macmillan Magazines Limited.)

methylated. Thus, it is possible that the reactivation of genes on the inactive X during ageing is due to loss of methylation, owing to a failure of the normal maintenance mechanisms. Reactivation of the human inactive X chromosome appears to be much less common during ageing than in the mouse, suggesting that the epigenetic controls are much tighter in long-lived animals (see Chapter 7 and Note 4.9). It will be easier to test the epigenetic theory of ageing when more information accumulates about the mechanisms that normally control the activities of specialised genes.

TOXICITY

Amongst the many theories of ageing that have been proposed this century, one of the earliest was that of Metchnikoff (1907), who proposed the production of various toxic products in the bowel was an important cause of ageing. These ideas are completely outdated, yet there is a substantial body of opinion that attributes at least some of the features of ageing to various types of environmental insult.

Some plant toxins are carcinogenic, or chemicals in plants can be converted into carcinogens or mutagens during the detoxification process. It is probable that the different incidence of particular cancers in various parts of the world is due, at least in part, to variation in diet (Cairns 1978). It would be surprising if toxic components of diet did not have other deleterious effects on cells and tissues, which contribute to ageing. For example, there are plant toxins that induce neurological disease (reviewed by Nunn 1994). It is at least possible that a 'perfect' diet, which lacked all toxins but contained all essential nutrients, would tend to increase lifespan. If this was the case, there would be some truth in the toxicity theory of ageing, but it cannot be said we age because we ingest toxins.

There are also other important environmental hazards. Smoking causes lung cancer and is damaging to the cardiovascular system. Smoke from wood fires can induce cataracts in third world countries (Balasubramanian, Bhat & Rao 1990). The ultraviolet light components of sunlight (UVA and UVB) can induce photoageing, that is, premature wrinkling of unpigmented skin, as well as basal cell carcinomas. There is ongoing debate about the possibility that aluminium is a cause of Alzheimer's disease. Other neurological diseases are produced by ingesting brain tissue, or other tissues, containing infective prion particles (see Note 4.10). This is an active field of research that may throw new light on the way complex macromolecules (distinguishable from pathogens) can induce degenerative changes in cells and tissues.

There is, of course, much public interest in the effects on the ageing process of diet, life style and so on, and part of this interest is in antioxidants or other treatments that may modulate ageing. Unfortunately, this is also an area where pseudoscience is influential and anecdotal evidence is all too commonly believed. Two areas that are being intensively studied by proper scientific investigation are the effects of calorie restriction on longevity, and the importance of oxygen free radicals and defences against free radical attack. Ultimately, these results will be related to the possible significance of diet, or other environmental effects, in modulating lifespan.

ENDOCRINE AND PROGRAMME THEORIES

Whereas most of the theories that have so far been considered are based on changes in macromolecules and cells, there is another possibility that relates more to physiological changes. Major homeostatic mechanisms depend on hormone signalling, particularly the neuroendocrine system. Wherever complex mechanisms exist, in which positive and negative feedback systems are important components, there is always the possibility of abnormal fluctuations that may have long-term consequences (Finch 1987, 1990). It is perhaps too difficult to formulate precise mechanisms along these lines, but the general idea of a pacemaker or clock in the central nervous system has often been discussed. Certainly we know that hypophysectomy (removal of the anterior pituitary) substantially increases the lifespan of rats, with physiological or biochemical effects comparable to calorie deprivation (Everitt et al. 1968, 1980).

The idea of a pacemaker also includes a whole host of proposals about 'programmed ageing'. For the most part these are ill defined, often in fact more a statement of opinion or belief rather than a precise proposal. If the programme is an intrinsic part of development, reproduction, a post-reproductive phase and final senescence, then defenders of the theory could maintain that we need to understand development first, before we can understand subsequent events. This is a reasonable argument, but it does not address the evolutionary problem, that is, why should a programme per se evolve? The easiest solution is to be aware of the evolved design of animals, which shows that a programme, in effect, resides in the design, and that the design is not compatible with indefinite survival. This viewpoint is discussed in Chapters 1, 2 and 8, whereas Chapter 6 considers more specific programmes, which occur in a minority of biological situations.

CONCLUSIONS

The theories of ageing discussed in this chapter all relate in one way or another to the maintenance mechanisms discussed in Chapter 3, and this is summarised in Table 4.3. Maintenance mechanisms have evolved to counter various intrinsic or extrinsic damaging events, and the theories considered mainly fall into the general category of stochastic, or wear and tear, theories of ageing. Most consider events in macromolecules and cells, but some are concerned with higher orders of organisation, such as the immunological or neuroendocrine theories. In this connection it is reasonable to distinguish those that at-

Table 4.3. *Correspondences between maintenance mechanisms (Chapter 3) and specific theories of ageing (Chapter 4)*

Maintenance mechanism	Theory of ageing	Primary references
1. DNA repair	Somatic mutation	Szilard (1959); Curtis (1966); Burnet (1974)
2. Fidelity of synthesis of macromolecules	Errors and feedback	Medvedev (1962); Orgel (1963)
3. Defence against oxygen free radicals	Free radical damage	Harman (1956, 1981); Ames (1983)
	Abnormal mitochondria	Linnane et al. (1989)
4. Removal of defective proteins	Altered proteins	Rothstein (1975, 1979)
5. Immune response	Autoimmune reactions	Walford (1969)
6. Detoxification	Damage from toxins and chemicals	Metchnikoff (1907)
7. Wound healing	Wear and tear	—
8. Epigenetic controls	Dysdifferentiation and ectopic gene expression	Cutler (1982); Holliday (1984a, 1987)
9. Homeostatic mechanisms	Neuroendocrine	Finch (1987)

tempt to establish primary causes of ageing, such as irreversible changes in DNA or proteins, and others that are secondary. Thus, the immunological theory is secondary, since it does not address the reasons for the decline in lymphocyte function, or the specificity of recognition. Also, if the accumulation of defective proteins in cells is due to a failure of normal proteolysis, what actually brings this about? Nevertheless, the distinction between primary and secondary is inevitably blurred, especially if we take a more global view. This broader view of theories of ageing, namely, that there is some truth in all of them, implies that the deleterious events that finally lead to ageing can occur in many types of molecule and can be brought about by many inducing events. It also implies that there are many contributing causes to ageing, and this theme will be discussed at greater length in the final chapter.

In the gerontological literature it is common to discuss the merits and failings of this or that theory (see, for example, Warner et al. 1987). It might be better to discuss the eventual failure of this or that maintenance mechanism. As we saw in Chapter 3, the maintenance processes are very well documented in an extensive literature, not much of which relates to ageing per se. So what is needed is a more thorough examination of the efficiency of maintenance in adults as they age, and also in species with different lifespans.

5

Cellular ageing

The finite lifespan of the soma also means that somatic cells cannot survive indefinitely. It is nevertheless possible that the death of somatic cells is not due to intrinsic cellular events, but rather to extrinsic ones. Their eventual death could be due to the failure of one or more of the interactions that are necessary for the normal functions of the organism. Such interactions can occur at a distance mediated, for example, by growth factors or hormones, or at the level of cell–cell contacts, within or between tissues. Although a breakdown of interactions between cells and tissues may well be an important component of ageing, there is much evidence that cells are subject to intrinsic changes that lead to their senescence and death. A very good example of the dramatic changes that can occur in cells as a result of normal ageing is shown in Figure 5.1a,b. The lens from an infant and an 80-year-old individual were obtained post-mortem, and the epithelial cells were then grown in primary culture under the same conditions. There is much experimental evidence that dividing cells undergo ageing both *in vitro* and *in vivo*. In the former situation, cells are studied in culture under well-defined conditions. In the latter, their fate is followed by transplantation from animal to animal.

THE FINITE LIFESPAN OF CELLS IN CULTURE

When Weismann realised that there is a crucial distinction between germ line and somatic cells, he also made the remarkable prediction that all somatic cells should also have finite lifespan (Kirkwood & Cremer 1982). For many years in this century it was believed that this prediction was incorrect. This belief came largely from the experiments of Carrel, which received much publicity at the time. He was a pioneer in the development of cell culture techniques, and his best-known experiments were with primary chicken cells. These cells were sequentially subcultured, and he claimed that they could be grown continuously for as long as 27 years, which was much longer than the lifespan of the donor organism. It is now known that Carrel's experiments were not carried out in the way he claimed (see Witkowski 1980, 1987), and also that primary chicken cells divide only 30–40 times before growth ceases.

The belief that cultured cells could be grown indefinitely was not only due

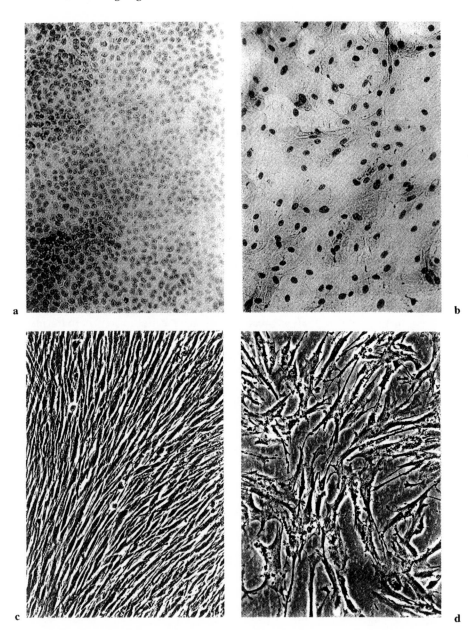

Figure 5.1. (a,b) Human lens epithelial cells obtained post-mortem and grown in primary culture: (a) from a 1-year-old infant, (b) from an 80-year-old individual. (c,d) Cultured human diploid fibroblasts, strain MRC-5: (c) a confluent layer of early passage cells; (d) senescent cells at late passage, which fail to become confluent. (Parts (a,b) courtesy of Y. Courtois.)

to the work of Carrel. When cell culture techniques became more standardised, cell lines were identified that could grow indefinitely. The most well known were HeLa cells, which were derived in the early 1940s from a tumour of the cervix from a woman. Various sublines of HeLa cells are widely used today in experimental studies. There are also rodent cell lines, such as the L cell line, that grow indefinitely. Both HeLa and L cells are transformed cells, that is, they have abnormal karyotypes and many morphological and other differences from primary diploid cells (see later in this section). Nevertheless, the view prevailed for many years that cultured mammalian cells grew indefinitely. In other terminology, they are referred to as permanent cell lines, or as being 'immortalised'.

In the late 1950s there were indications that primary diploid cells would not grow indefinitely, but usually it was assumed that the failure of growth was due to faulty medium, or some other unknown problem with culture conditions. Hayflick & Moorhead (1961) were the first to document fully the finite growth of human diploid fibroblasts, in which they defined three phases:

Phase I was the establishment of the primary culture from normal tissue (usually either foetal tissue, infant foreskin or a small skin biopsy).
Phase II was a long period of normal growth, in which the cell division time and the number of cells per flask was roughly constant.
Phase III was a period of growth when the cells began to grow more slowly, the yield of cells per flask declined, until finally the cells failed to reach confluence.

The end of Phase III is usually reached after 50–70 population doublings (PDs) from the initial Phase I culture. An example of a cumulative growth curve is shown in Figure 5.2 (see also Note 5.1). Although immortalised cells would be strongly selected in Phase III populations, such cells were not seen by Hayflick and Moorhead, nor have they been seen in innumerable experiments carried out in many laboratories subsequently. It is evident that spontaneous transformation or immortalisation of human fibroblasts either never occurs, or is an extremely rare event. In the later experiments of Hayflick (1965), the *in vitro* senescence of human diploid cells was more thoroughly documented. A number of controls established that the results were not due to defective medium or serum, or to an infecting virus or mycoplasma. It was shown, by mixing cultures of early passage male cells (XY chromosomes) and late passage female cells (XX), that the late passage cells died out first, and the male cells survived to the normal number of population doublings. It was also shown that the lifespan depended primarily on the number of cell divisions completed, rather than elapsed or chronological time. It was demonstrated that cells frozen in liquid nitrogen retained viability for many years and, possibly indefinitely, the growth potential they had when they were initially frozen. In spite of the very thorough documentation of the defined lifespan of these human diploid cells (reviewed by Hayflick 1977, 1980), many cell biologists of the time, and some to this day, refuse to believe that the

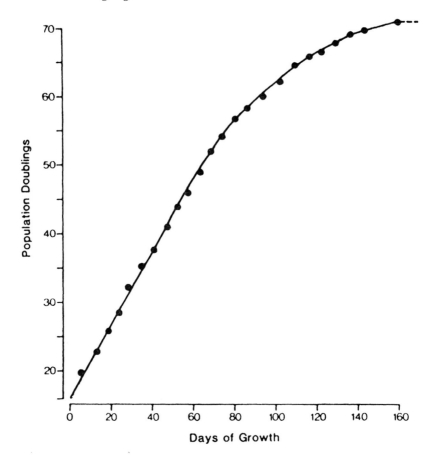

Figure 5.2. The cumulative growth curve of cultured human fibroblasts, strain MRC-5. Cells were subcultured as soon as they became confluent (1 : 8, 1 : 4 or 1 : 2 split ratios), and the population doublings (PDs) calculated from the yield of cells at each split. A long period of constant growth is superseded by a reduced growth rate, and finally no cell increase (see also Fig. for Note 5.3c).

'Hayflick limit' to cell growth is anything more than an experimental artefact, and that these cells would grow indefinitely if the correct conditions of growth were used.

Hayflick drew two important conclusions from his experiments. First, he proposed that the cessation of cell growth in Phase III was an example of intrinsic cellular senescence. These cells are morphologically quite distinct from young cells, as is shown in Fig. 5.1c,d. They do not line up in 'whorls' or parallel arrays, which is a characteristic of fibroblasts; they are variable in size; they may have abnormal lobed nuclei, or giant nuclei; they are generally more granular in appearance, and cells gradually detach from the substrate and form debris in the medium. Hayflick concluded that the eventual senescence

Table 5.1. *Human or bovine diploid somatic cells that have finite proliferation when cultured* in vitro

Cell type	No. of PDs	References
Fibroblasts	50–100	Hayflick & Moorhead (1961) and innumerable subsequent publications
Glial cells	20–30	Ponten & MacIntyre (1968)
T lymphocytes	60–170[a]	McCarron et al. (1987)
Smooth muscle cells	14	Bierman (1978)
Endothelial cells[b]	18	Mueller, Rosen & Levine (1980)
Keratinocytes	35–40	Rockwell, Johnson & Sibatani (1987)
Adrenocortical cells[b]	55–65	Hornsby & Gill (1978); Hornsby, Simonian & Gill (1979)
Thyrocytes	unmeasured	Davies et al. (1985)
Bronchial epithelial cells	35	Lechner et al. (1981)

[a]The authors report an expansion of 10^{19}–10^{52} for mass cultures and 10^4–10^{35} for individual clones.

[b]Bovine cells; all others are human. It is reported in a less quantitative study that human endothelial cells grow for 25–35 population doublings (PDs).

of these cells provided an excellent experimental opportunity for the further study of ageing at the cellular and molecular level. Second, he made the crucial distinction between cell strains that retain a diploid karyotype and had finite growth in culture, and transformed lines, which have an abnormal heteroploid karyotype and can grow indefinitely in culture. Diploid fibroblast strains form a confluent monolayer, that is, they are subject to 'contact inhibition' and are anchorage dependent. Cell lines on the other hand normally continue to grow when confluent and are not anchorage dependent, that is, they can grow to form colonies in soft agar, or cell suspensions in liquid medium. Unfortunately, Hayflick's distinction between lines and strains has not been followed by later workers, and this has led to much confusion. Part of this confusion is due to the existence of cells that do not readily fall into one category or the other. For example, 3T3 mouse cells are immortalised, but they retain a fibroblast morphology and are subject to contact inhibition (Todaro & Green 1963). They are frequently used to study the process of transformation in which cells form foci, lose contact inhibition, are not anchorage dependent and are often tumorigenic when injected into nude mice. In reality, 3T3 cells have undergone at least one step on the road to full transformation and, in Hayflick's terminology, they constitute a permanent cell line.

The observations of Hayflick and Moorhead with human fibroblasts have been repeated by many other investigators. It has also been shown that fibroblasts from other species have a finite lifespan. Other cell types have been shown to have finite growth in culture as well. These include glial cells, keratinocytes, epithelial cells, endothelial cells, T lymphocytes, muscle cells, thyrocytes and cells from the adrenal cortex. Table 5.1 summarises these results.

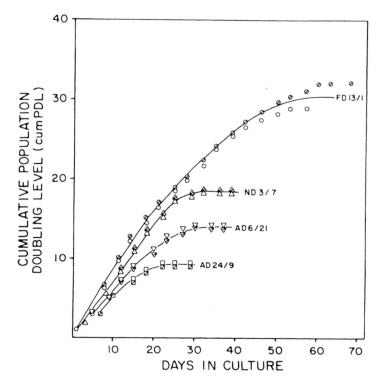

Figure 5.3. The cumulative growth of Syrian hamster dermal fibroblasts obtained from animals of increasing age. FDB/1, 13 days' gestation; ND3/7, 3-day-old neonatal; AD6/21, 6-month-old young adult; AD24/9, 24-month-old adult. (Reproduced with permission from Bruce, Scott & Ts'o 1986.)

There is no known example of primary diploid cells that will grow indefinitely in culture. This is a remarkable situation, since the cells that are being studied have very different phenotypes and very different roles in the organism, yet in every case they seem to die eventually from an intrinsic ageing process. In addition, it has been shown that cells from young donors have a greater division potential in culture than cells from old donors. Hayflick (1965) published the first evidence, and it was subsequently confirmed by more comprehensive studies (Martin et al. 1970). An even stronger relationship between donor age and lifespan in culture has also been shown for hamster fibroblasts (Fig. 5.3). Human T lymphocytes have been obtained from individuals of different age and grown *in vitro* until cell division ceased. Although there was some relationship between donor age and lifespan *in vitro*, this was largely obscured by the extreme variation in the division potential of both populations and individual clones (McCarron et al. 1987).

Another very important approach involves the comparative study of the life-

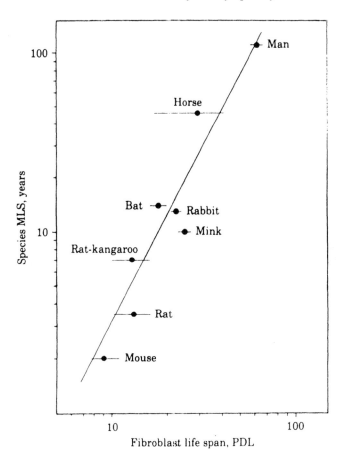

Figure 5.4. The log of the lifespan of cultured fibroblasts in population doublings (PDL), and the log of the lifespan of donor species. (Reproduced with permission from Rohme 1981.)

span of fibroblasts from species with different longevities. Fibroblasts from mouse embryos grow for only 10–15 population doublings before they become senescent. (However, these populations, unlike human cells, often become immortalised, so that growth continues after a lag phase.) Chicken and hamster fibroblasts have a growth potential intermediate between those of human and mouse. A definitive study was carried out by Rohme (1981) using eight mammalian species. His results are shown in Fig. 5.4, and they demonstrate a clear correlation between donor lifespan and the growth potential of their cells. Finally, it has been shown that fibroblasts from patients with the premature ageing disease Werner's syndrome (see Chapter 8) have very limited growth *in vitro* (Martin et al. 1970; Thompson & Holliday 1983).

All these features of cellular ageing indicate that the intrinsic changes seen relate in some way to the ageing of the organism. The results do not demonstrate that organisms age because cells capable of cell division are becoming senescent. What they do appear to demonstrate is that various cells of the body that can divide *in vivo* as well as *in vitro* have some intrinsic property that limits their indefinite survival. It would be very surprising if this property had no relationship to the ageing of the whole organism.

TRANSPLANTATION EXPERIMENTS

The survival of cells *in vivo* has been studied by transplanting cells or tissues from animal to animal. Inbred mice have the same histocompatibility antigens, so they will accept skin grafts from another animal. Some of the earliest transplantation experiments were carried out using mouse skin (Krohn 1962). He demonstrated that the same piece of skin could survive on two or more successive recipients for 6–7 years, which is longer than the lifespan of each animal. Nevertheless, the skin eventually showed the usual features of senescence, and reached a point where continued transplantation was not possible. Similar experiments were carried out later on mouse mammary tissue (Daniel et al. 1968; Daniel and Young 1971). Tissue from young females was transplanted to recipients from which the same tissue had been surgically removed. The transplant grew to form perfectly normal mammary glands. This animal was then used as a donor for a further transplantation experiment. It was found that, after several transfers, the growth of the mammary tissue declined and functional gland tissue was no longer produced. This result was in complete contrast to the transplantation of mammary tumour tissue using the same techniques. The tumour tissue could be propagated from animal to animal indefinitely because the transformed cells in the tumour were immortalised.

Many experiments have been carried out on the sequential transplantation of the haematopoietic cells of the bone marrow. In these experiments, the recipients are either an anaemic strain that cannot generate its own population of bone marrow cells, or are heavily irradiated to destroy the pre-existing population of stem cells. In many experiments several transplant generations have been possible over a period longer than the lifespan of a mouse, but in all cases the populations of stem cells eventually died out. Whether this means that haematopoietic stem cells have finite lifespan is a contentious issue, one that has been fully reviewed and discussed many times (see, for example, Harrison 1979, 1984, 1985; Harrison et al. 1989, and also Note 5.2). An important study has been carried out with a specific cell type. Mice were immunised to a particular antigen, and the specific antibodies produced by memory cells were identified (Williamson & Askonas 1972). These cells are capable of division and retain a memory of the particular antigen; thus, a later dose of antigen results in their proliferation and production of antibody. The cells can be transplanted to an irradiated recipient of the same inbred strain, and will then divide in response to antigenic stimulation. It was found that the transplantation of these cells was initially successful, but after a few transplant generations antibodies could no longer be detected after injection of antigen. This result shows that the cell population had died out. It was calculated that the total number of cell divisions was in the region of 100, which is not very different from the Hayflick limit. In other transplantation experiments, bone marrow cells were taken from young and old donors. Their capacity to proliferate in recipient animals was not different, which indicates there is little, if any, ageing of these cells in their natural environment (Ogden & Micklem 1976; Harrison et al. 1989).

All these results indicate that normal cells *in vivo* have limited ability to proliferate, but the extent of proliferation is probably considerably greater than would be expected during the normal life time of the animal. It seems that cells have greater survival time than is necessary for the organism, but this may be just part of a homeostatic mechanism that gives inbuilt security. To ensure survival under abnormal conditions, organisms have evolved safety measures, so that if one physiological process or mechanism fails, there is often a back-up process to take its place. The ability of an organism to survive adverse treatment is shown by irradiation experiments. At an appropriate dose a large proportion of the haematopoietic stem cells are killed, but some survive. These cells can renew the whole population of haematopoietic cells. It has not been established how often sequential renewal can occur, because successive irradiations are liable to induce leukaemias, or other abnormalities in the animals, which complicates the interpretation of the results.

STEM CELLS AND COMMITMENT TO SENESCENCE

Cells that continuously divide in somatic tissue are usually stem cells. These are outwardly undifferentiated, but when they divide they can produce daughter cells that are destined to differentiate, either immediately or after some further cell divisions. Some of the daughter cells retain the stem cell phenotype. Examples of stem cells are those that replenish the epidermis of the skin, the cells at the base of the villi of the gut, and the various cell types in the blood. In the latter case, many types of differentiated cell are produced from one stem cell population. Stem cells may divide every 20 hours or so continuously throughout the lifespan of the animal, so the question of their maximum proliferative potential has often been raised. This is, however, difficult to determine by experiment. One method depends on the labelling of DNA with ^{3}H-thymidine. If a stem cell population divides continuously, then all the cells become labelled; but if a cell stops dividing, then unlabelled cells should be detected. Those who carry out such studies claim that unlabelled cells are either never seen, or are extremely rarely seen, which suggests that the cells can divide continuously many hundreds of times, and perhaps indefinitely (Potten & Loeffler 1990).

A simple calculation shows how hard it is to be certain about the sequential number of divisions in a given cell lineage. Suppose there is a quiescent or semi-quiescent pool from which stem cells are drawn, and that each cell then divides about a hundred times. Only ten cells are then needed to produce a thousand sequential divisions. It is also debatable whether the autoradiographic procedure can indeed detect the occasional unlabelled cell. How would one know that any unlabelled cell seen (and there would be many in the tissue being studied) is not a non-cycling or senescent stem cell? The behaviour of skin keratinocytes in culture illustrates the complexity of the problem. These cells can be grown in two ways, depending on the amount of calcium in the medium (Rockwell, Johnson & Sibatani 1987). One pathway is tissue differentiation, in which the cells divide a few times and then cease division and produce very large amounts of keratin. Another pathway is simply proliferation without differentiation: these cells divide 20–40 times and become terminally senescent, without differentiating. It is clear that a single cell can produce large numbers of undifferentiated offspring that are like stem cells, since each of these, if placed in the appropriate conditions, would divide a few times to produce differentiated cells. The overall growth potential of a relatively small founder population of keratinocytes is enormous.

The organism develops from germ line cells and transmits its own potentially immortal germ cells to the next generation. At some stage in development, therefore, cells change from their immortal phenotype to one with finite growth. In other words, cells become committed to senescence, although the overt signs of senescence might occur very much later on. This commitment might occur in the developing embryo as soon as the somatic cells are segregated from the germ cells. Alternatively, commitment could occur much later on in development, and some believe that uncommitted, immortal cells exist in the adult.

The concept of commitment is useful in trying to understand the difference between populations of cells that die out and those that keep growing indefinitely. More specifically, it can be applied to populations of human fibroblasts (Kirkwood & Holliday 1975b; Holliday et al. 1977; Holliday, Huschtscha & Kirkwood 1981). The commitment theory proposes that there is a population of uncommitted, immortal cells that have a given probability P of producing a committed cell by division. The committed cell continues to divide, and eventually all its progeny become senescent and stop dividing. The number of divisions from commitment to the non-dividing state, the incubation period, is designated M. The third important parameter is N, the population size. It can be shown that if P is 0.25 or greater and M is 55 or more, then the populations of the size normally grown in the laboratory (10^6–10^7) will invariably die out, as all the uncommitted cells will be lost by simple dilution (see Note 5.3). The model is important because it may also explain the failure to repopulate bone marrow cells in the serial transplant experiments in mice described above (see 'Transplantation experiments'; also Notes 2.2 and 5.2). If there is a small subset of immortal uncommitted cells, then these will be lost by dilution during the course of serial transplantation, and once this happens the whole population can never be replenished. The significance of population size N was demonstrated experimentally. In primary cultures where a small proportion of uncommitted cells may exist, a reduction in population size (a 'bottleneck') may have no effect on final lifespan; but if the bottleneck contains no uncommitted cells, then the final lifespan is significantly reduced (see Note 5.3).

In the fibroblast model it is easy to show that reducing the value of either P (<0.25) or M (<55), or increasing N, can convert a population with finite growth potential to one with infinite growth potential. This also highlights the important difference between the ageing of whole populations and the ageing of the individual cells and cell lineages (see Chapter 1). The commitment model also explains some of the known quantitative features of fibroblast populations. Parallel populations of the same strain started at an early passage level do not have the same longevity in terms of population doublings, but vary over a wide range (Holliday et al. 1977). Also, individual clones are known to have very different longevities (Smith, Pereira-Smith & Schneider 1978; Smith & Whitney 1980). Late passage or senescent populations would often be totally derived from the longest-lived clone, as has been shown to be the case with cultured T lymphocytes (McCarron et al. 1987).

In conclusion, it is clear that there are many somatic cells with finite proliferative potential, but it is not certain whether or not there are subsets of cells with infinite growth potential. The experimental analysis of populations of cultured cells is hard enough; to understand the growth potential and ageing of dividing cells *in vivo* is of much greater difficulty.

MECHANISMS OF CELLULAR AGEING

In attempting to understand the reasons for the finite growth of mammalian diploid cells in culture, it is essential to realise that events must be occurring throughout the Phase II period of apparently normal growth. This is true of all types of ageing: as time proceeds, the time remaining prior to senescence and death decreases. The major difference in the case of the ageing of human fibroblasts is that the important parameter is population doublings, rather than chronological time. The events that might be occurring fall into two general categories. The first entails that cellular ageing be regulated or programmed, which is assumed by many investigators in the field; however, in most publications the way in which the programme might operate is not specified. It is possible to invoke a cell division counting mechanism based on a given number of repeats of a given stretch of DNA (see Note 5.4), but no evidence of such a molecular clock mechanism exists. The second broad possibility is that ageing of diploid cells is a 'multiple-hit' process. During sequential divisions, stochastic or random molecular defects might occur that are transmitted to the next generation, so that they gradually accumulate. Initially the phenotype is unchanged, but as more and more defects occur, cell division is impaired, senescent changes are seen and finally growth ceases. One can think of this process as an example of unbalanced growth. Suppose only a proportion of defects are removed or repaired at each division; then there will still be a residue transmitted to daughter cells. The accumulation of defects would have to proceed inexorably, because all the cells of the population eventually die out. In other words, there can be no opportunity for the continual selection of cells with few defects. According to the theory, either cells that are immortalised have achieved an equilibrium between the occurrence of defects and their removal, so that they are in a steady state, or there is continual selection in the population for cells with the fewest defects. The 'multiple-hit' explanation of cellular ageing becomes particularly attractive if the increase in defects is not linear with sequential cell division, but is instead exponential. Any feedback mechanism (see Chapter 4 and Fig. 4.2) will give such an exponential increase in errors, so that the phenotypic effects will become more extreme at the end of the lifespan. Many of the changes which are seen during fibroblast ageing do appear to accelerate in Phase III (see below).

There is one attractive possibility that encompasses both a counting mechanism and stochastic events. There is a problem in the replication of linear DNA, because without a special molecular mechanism, some DNA will be lost at each round of replication (see Note 5.5). The special mechanism depends on the existence of repeated DNA telomeric sequences and the enzyme telomerase. In the absence of the enzyme, telomeric sequences are lost and eventually genes proximal to these sequences will also be lost. Some of the genes will have essential functions, so there will be progressively deleterious consequences to the cell. It was suggested many years ago that the progressive loss of terminal chromosomal DNA might be a cause of ageing (Olovnikov 1973). There is now evidence that the amount of telomeric DNA does indeed decline during the ageing of human fibroblasts, whereas permanent lines have a characteristic amount that stays constant (Harley, Futcher & Greider 1990; Harley 1991; Counter et al. 1992). Also, the growth potential of skin fibroblasts from donors of different age correlates well with the amount of telomeric DNA in the initial cell population (Allsopp et al. 1992). Commitment to senescence could then be due to the loss of telomerase from the cells, and immortalisation would depend on its reacquisition. The loss of telomeric DNA could be regarded as a molecular clock, but it could also be stochastic because the number of telomeric repeats may vary between chromosomes in individual cells, as well as between different cells. As loss of telomeric DNA proceeds, the phenotypic consequences could be quite variable, depending on the heterozygous or homozygous loss of important genes near the ends of chromosomes.

In one of his reviews, Hayflick (1980) lists no less than 167 physiological or biochemical parameters that have been examined, and of these 117 change during senescence. This body of data strongly indicates that there are pleiotropic or multiple effects of senescence. As previously mentioned, the cells in Phase III are themselves heterogeneous in size and morphology, and electron microscope studies demonstrate that the granularity seen under phase contrast is due to inclusions and secondary lysosomes. Chromosome studies demonstrate an increase in abnormalities in senescent populations, including aneuploidy and polyploidy. It is also likely that gene mutations increase exponentially during ageing (see Note 5.6). There is a progressive loss of 5-methyl cytosine in genomic DNA (Wilson & Jones 1983; Fairweather et al. 1987), which might influence ectopic gene expression. There is also evidence for the accumulation of defective proteins in senescent fibroblasts. As we saw in Chapter 4, proteins are subject to errors during synthesis and also to post-synthetic changes, and defective molecules can be removed by proteases. If there is not a balance between the formation and removal of defective proteins, then the cells would not be in a steady state. There could then be a gradual accumulation of altered molecules, which the cells could tolerate for a long time, but which ultimately would produce the senescent phenotype. It is also possible that there could be a more rapid build-up of defects towards the end of the lifespan, by feedback of errors into the pathways for protein synthesis. The general hypothesis is attractive because it can certainly, in principle, explain the changes in the parameters reviewed by Hayflick, since proteins are involved in so many cellular activities (see also Chapter 9 and Fig. 9.1). Major phenotypic changes that occur during the ageing of human fibroblasts are listed in Table 5.2.

Although there is much evidence for various kinds of biochemical, genetic or physiological defects in senescent fibroblasts, it is not widely believed that such defects constitute a primary cause of the ageing of these cells. Instead, it is more commonly assumed that senescence of human fibroblasts is in some way programmed or regulated. A popular hypothesis is that the eventual failure of the cells to divide is due to a block in the initiation of DNA synthesis, either because there is an endogenous inhibitor, or because they become unresponsive to the growth factors in serum that are mitogenic for young cells (see, for example, Smith 1984; Smith & Lincoln 1984; Stein & Atkins 1986; Norwood, Smith & Stein 1990; Smith 1990). The belief that senescence is regulated depends largely on the finding that these cells have some dominant or positive function, which can impose a senescent phenotype in hybrids or heterokaryons. Thus, hybrids between diploid cells and immortalised cells will most commonly undergo senescence (Bunn & Tarrant 1980; Pereira-Smith & Smith 1983). Also, heterokaryons between senescent cells and young cells result in the cessation of DNA synthesis in the young nuclei (Norwood et al. 1974). These results have been interpreted to mean that senescent cells produce an inhibitor of DNA synthesis, and that this function is

Table 5.2. *Some important phenotypic features of senescent human fibroblasts (see also Hayflick [1980] for a detailed compilation)*

Phenotype	Representative references
1. Cells do not line up in parallel arrays, are variable in size and do not become confluent	Hayflick & Moorhead (1961); Hayflick (1965)
2. Increasing granularity, secondary lysosomes and autofluorescence	Robbins, Levine & Eagle (1970); Lipetz & Cristofalo (1972); Jongkind et al. (1982); Rattan et al. (1982)
3. Chromosome abnormalities, aneuploidy and polyploidy	Saksela & Moorhead (1963); Thompson & Holliday (1975)
4. Reduction in ^{3}H thymidine labelling; cessation of DNA synthesis	Cristofalo & Sharf (1973); Vincent & Huang (1976); Holliday et al. (1981)
5. Reduced rate of replicon elongation	Petes et al. (1974)
6. Reduction in telomeric DNA	Harley et al. (1990); Levy et al. (1992)
7. Reduced fidelity of DNA polymerase	Linn, Kairis & Holliday (1976); Murray & Holliday (1981)
8. Increased frequency of G6PD variants	Fulder & Holliday (1975)
9. Loss of cytosine methylation in DNA	Wilson & Jones (1983); Fairweather et al. (1987)
10. Alterations in protein turnover	Bradley, Hayflick & Schimke (1976); Shakespeare & Buchanan (1976, 1979)
11. Appearance of heat labile G6PD, 6PGD and TPI	Holliday & Tarrant (1972); Tollefsbol et al. (1982)
12. Increased lysosomal enzymes	Cristofalo, Parris & Kritchevsky (1967); Wang et al. (1970); Milisauskas & Rose (1973); Cristofalo & Kabakjian (1975)
13. Increased absorption of conconavalin A and RBC	Aizawa & Mitsui (1979)
14. Reduced fidelity of protein synthesis	Luce & Bunn (1989)
15. Increased sensitivity to paromomycin	Holliday & Rattan (1984)
16. Delay in induction of ODC	Fulder & Tarrant (1975)
17. Abnormalities in fibronectin	Chandrasekhar et al. (1983); Edick & Millis (1984)

Abbreviations: G6PD, glucose-6-phosphate dehydrogenase; 6PGD, 6-phosphogluconate dehydrogenase; TPI, triosephosphate dehydrogenase; ODC, ornithine decarboxylase; RBC, red blood cells.

absent in immortalised cells. (It should be noted, however, that cells containing molecular defects, when fused to cells largely free of such defects, could 'contaminate' the recipients; in other words, random errors could also have a dominant effect [see also Note 5.7].) If the lifespan of fibroblasts is regulated or programmed, then the definition of senescence becomes a central issue, because senescence implies degenerative change, loss of homeostatic mechanisms and so on. A programmed event that results in the inhibition of the G1–S transition is a different situation, and does not by itself constitute senescence (see Note 5.8). It calls into question Hayflick's view that Phase III cells provide experimental material for the study of ageing at the cellular level. Moreover the programme theory does not concern itself with events during Phase II, apart from the suggestion that a repressor of the inhibitor of DNA synthesis is coded for by many gene copies, which are progressively lost during serial subculture (Smith & Lumpkin 1980). It is also commonly stated that Phase III cells resemble quiescent cells, in that they are blocked in G1, and that they can be held in this quasi-quiescent or senescent state for long periods of time, provided the medium is changed weekly. This is not in accord with extensive documentation, including Hayflick & Moorhead's (1961) original studies of heterogeneity in senescent populations. In the experience of many investigators, the final population of non-dividing cells becomes increasingly abnormal, and since cells disintegrate or detach from the surface, the number remaining constantly declines. This clearly substantiates Hayflick's claim that human fibroblasts are indeed an appropriate experimental system to study ageing at the cellular level.

Although it is widely believed that Phase III is due to a dominant regulated function, it is far from clear why such a function should exist. The only serious possibility that has been proposed is that senescence is in some way a barrier that prevents the emergence of cancer cells (Dykhuizen 1974). It is true that many cancer cells have been grown indefinitely, that is, they have become immortalised. It is also true that normal fibroblasts can be transformed by certain viruses to abnormal cells, like neoplastic cells. Yet these are not immortalised and they are not tumorigenic. The behaviour of these cells suggests that a barrier does exist that prevents the emergence of cancer cells. In contrast, human B lymphocytes can be readily immortalised by the Epstein–Barr virus, but these cells appear to be quite normal. Indeed, many individuals who have contracted glandular fever have such cells in their blood, with no deleterious effects. The interrelationships between senescence, transformation and immortalisation are complex, and are discussed in more detail in the next section.

It should be borne in mind that the growth potential of individual human cells is extremely large ($2^{60} = 10^{18}$ cells, with a net weight of about 10^6 kg), so any malignant cancer cell could expand to large populations, easily sufficient to kill the individual, long before the 'barrier' was reached. It is well known that avian cells can be transformed by viruses, but in almost all cases these cells do not constitute a permanent line. Nevertheless they are malignant when injected into a chicken, and the animal is usually killed by their proliferation. For this reason alone, it seems unlikely that the biological basis of senescence is to provide a barrier to the emergence of cancer.

An important attempt has been made to reconcile the stochastic and regulatory hypotheses (Rosenberger, Gounaris & Kolettas 1991). It is well known that bacteria and eukaryotes respond to environmentally-induced stress. Thus, heat shock or UV treatment results in the synthesis of many proteins that collectively help the cell to survive, and during the response, major activities such as transcription and DNA replication may be shut down. It is therefore possible that during the growth of cells in Phase II, defects are indeed occurring in macromolecules, but that when these reach a certain level, the cell responds by turning on a stress response. This in turn results in the inhibition of DNA synthesis and cell division, but in this case without recovery.

In conclusion, one can state with certainty that experimental studies have established the reality of intrinsic ageing of individual cells and populations of cells. There are also strong grounds for believing that the growth potential of cells is related to the maximum lifespan of the species. Primary diploid fibroblasts grown in culture provide excellent experimental material for the study of the mechanism of cellular ageing, and innumerable investigations have been carried out. Nevertheless, the underlying mechanism remains very elusive, and this poses a challenge for future investigators.

THE IMMORTALISATION OF SOMATIC CELLS

The intrinsic ageing of dividing somatic cells might become clearer if the process of immortalisation, or escape from senescence, was understood. It has been known for a long time that primary mouse or rat fibroblasts that become senescent in culture will spontaneously produce immortalised derivatives. These cells are selected from large populations of senescent or non-growing cells. They have an apparently normal morphology (e.g. 3T3 cells), but they are permanent lines with unlimited growth potential. These cells can now be transformed by a variety of carcinogens to cells with completely altered morphology, which can be tumorigenic in nude mice. It is known that these events can depend on mutations in oncogenes (Land, Parada & Weinberg 1983). Also, transforming oncogenes may be present in viruses. Although activated or mutant oncogenes can immortalise and transform rodent cells, the equivalent genes do not have the same effect on human somatic cells. The fact that the senescence of these is a reproducible observation by itself means that immortalised cells are not being formed, because if any were produced they would keep growing and form a permanent cell line. Moreover, a variety of mutagenic or carcinogenic treatments have failed to immortalise these human cells.

Experiments have been done in which oncogenes that are known to immortalise or transform rodent cells have been transfected into normal human cells (usually by using a retroviral vector); however, the cells are neither transformed or immortalised. There are other genes that can produce these changes, but these are in the genomes of viruses. When human fibroblasts are infected with simian virus SV40, the virus does not proliferate and kill the cells, but it transforms them to a new morphological type, with the characteristics of neoplastic cells. These cells have an epithelial morphology, do not show contact inhibition, and can divide when not attached to a solid substrate. They also have a reduced requirement for growth factors in serum, and an abnormal karyotype. The virus gene responsible for these changes codes for a protein known as 'large T antigen'. Although these cells have a transformed phenotype, they are not immortalised. They may grow for a longer time than the cells from which they were derived, but eventually growth ceases, and the cells are said to enter 'crisis'. It is a matter for debate whether crisis is the same phenomenon as the senescence of normal cells (see Note 5.9).

What is surprising is that the SV40-transformed cells only very occasionally produce immortalised derivatives (Huschtscha & Holliday 1983). This strongly indicates that a second rare event is necessary for immortalisation. In this regard, human fibroblasts are completely different from rodent cells. However, other types of human cells can be much more readily immortalised by virus genes. As previously mentioned, the Epstein–Barr virus can immortalise human B lymphocytes, and DNA from the papilloma virus can immortalise human epithelial and other cells (reviewed by DiPaulo et al. 1993).

As well as oncogenes, there are also antioncogenes or tumour suppressor genes (see Shay, Wright & Werbin 1991; Levine 1993). Many tumour cells have mutations in the tumour suppressor gene that codes for the p53 protein. It is known that the large T antigen interacts with the normal non-mutant p53 and alters its properties. This is responsible for the transformed phenotype of SV40-infected cells. However, some other change is required for immortalisation and full transformation, which could be a mutation in p53 or some other tumour suppressor gene, or the activation of telomerase (see Note 5.9).

Although all these results are of considerable interest, they have not yet explained the molecular basis of either cell senescence or immortalisation. The clearest conclusion that one can draw is that human cells are very refractory to transformation and immortalisation in comparison to mouse or rat cells, and that cells from some other species, such as Syrian hamster, are intermediate. This result supports the view that the phenotype of specialised cells is maintained much more effectively in long-lived species than in short-lived ones. This maintenance may depend on the epigenetic controls of gene activity (see also Chapters 3 and 7).

NON-DIVIDING CELLS

The discussion in the previous sections of this chapter has been primarily concerned with populations of dividing cells, but many tissues of the body consist largely or entirely of post-mitotic, non-dividing cells. These are usually highly specialised cells, such as neurones and muscle cells, and they are also long lived. In some cases the death of a cell by an intrinsic cause, or an externally induced injury, is followed by replacement, but in other cases this does not happen. Thus, damaged muscle (other than cardiac muscle) can be replaced by the subsequent growth and differentiation of myoblasts, which are present in muscle tissue. The loss of a neurone, on the other hand, is not followed by its replacement. Most gerontologists would agree that the death of non-dividing cells is a very important component of the ageing of the whole organism.

It is important to try and compare the ageing of dividing cells, as discussed above, with that of non-dividing cells. One major difference is sensitivity of the DNA. It is well known that post-mitotic cells are not obviously damaged by doses of radiation that kill dividing cells. The reason for this is that induced double-strand breaks will not usually affect a cell that never enters mitosis, whereas such breaks can lead to loss of genetic material at mitosis. The irradiation of PHA-stimulated lymphocytes results in the formation of micronuclei (containing a chromosome or chromosome fragment), but micronuclei are not formed by irradiated post-mitotic cells. Nevertheless, post-mitotic cells have effective DNA repair mechanisms, because the continual spontaneous damage (such as the deamination of cytosine to uracil or depurination) would otherwise interfere with normal transcription, or with the regulation of gene activity.

Another major feature of specialised non-dividing cells is the absence of cellular selection. When cells are dividing dead cells may be produced, but there is always selection for viable ones; indeed, the fastest growing will always be the most strongly selected. In contrast, when a non-dividing cell dies, the event is terminal. With regard to molecular mechanisms of cell death, this has important implications. For example, if a lethal mutation occurs in one of two daughter cells, the other survives and can divide again. A lethal mutation in a neurone produces, by definition, only a dead cell. Also, the kinetics of the accumulation of deleterious recessive or codominant mutations will be different (see Note 5.10). It is probable that more mutations occur during DNA replication than in non-dividing DNA, but the higher mutation rate in dividing cells can be offset by continual selection, and the lower rate in non-dividing cells is simply additive. Very little information is available about the importance of mutations, or other DNA damage, in the ageing of non-dividing cells.

An escalating 'error catastrophe' in a non-dividing cell active in protein synthesis is irreversible and will kill the cell. In dividing cells, the most error-free cells will be selected, so the escalation of errors might be prevented (see Chapter 4). In contrast, dividing cells lose DNA telomere sequences, which eventually may be disastrous (see above, 'Mechanisms of cellular ageing', and Chapter 4), but non-dividing cells will retain their telomeres.

The situation is different for mitochondrial DNA. If oxygen free radicals damage mtDNA and the organelle loses function, there are two possible consequences. First, the defective mtDNA may still replicate, possibly faster than normal mtDNA, if the damage has deleted part of the genome. This replication of damaged mitochondria in a non-dividing cell is potentially disastrous, but in dividing cells, cells deficient in respiration will be selected against, so survival is possible. Second, a defective mitochondrion that cannot replicate is likely to be degraded, or replaced by the division of a normal organelle. In this case the effect is no different in dividing or non-dividing cells.

A likely cause of death of non-dividing cells is the accumulation of metabolic products that cannot be eliminated. In Alzheimer's disease amyloid deposits accumulate in brain cells, and the same histological changes are seen in older individuals in whom the disease has not been diagnosed. The continual turnover of photoreceptors in the rods and cones of the eye can lead to the accumulation of insoluble degradation productions in secondary lysosomes, which are phagocytosed by the underlying epithelial cells. These cells eventually lose normal structure and function, because they cannot maintain a steady physiological state (see Chapter 2). Similarly lipofuscin, which is thought to be a product of lipid peroxidation, accumulates during ageing in many non-dividing cells. Dividing cells, however, are not subject to the same accumulation, because the material is continually being diluted out.

A simple and elegant experiment was performed by Ponten and his colleagues (Brunk et al. 1973; Ponten 1973) using cultured human diploid cells. An individual cell can form a fairly large colony in which the chronologically oldest cells are in the centre and the youngest on the outside. The central cells accumulate lipofuscin and secondary lysosomes, whereas the outer ones do not. If half the colony is cut away with a scalpel, then the central cells proliferate more rapidly than the outer ones, because the former are younger in terms of elapsed cell divisions and the latter are older. Thus, the central cells are old in terms of their appearance, but young in terms of their proliferative potential, and the reverse is true for the outer cells. Also, when the central cells proliferate, the lipofuscin and secondary lysosomes are rapidly deleted out. These results indicate the difficulties in establishing definitive criteria for 'young' and 'old' cells (see also Note 5.11).

In general, the survival of a non-dividing cell will depend on its powers of replacement of components that may be damaged or lost. These components may be membranes, organelles, the endoplasmic reticulum, DNA and so on. We know that many neurones last as long as the lifetime of the organism, so these or other non-dividing cells of long-lived animals must have very effective cellular maintenance mechanisms. We would expect such mechanisms to be less effective in short-lived animals. Most comparative studies that have been done with cells or tissues from mammals with different longevities indicate that this is the case, and these are reviewed in Chapter 7.

APOPTOSIS

Recently there has been much interest in a cell suicide mechanism known as apoptosis. In many publications cell death by apoptosis is contrasted with cell death by necrosis. 'Necrosis' is a term used by pathologists to describe the terminal events that are seen when a cell has already died from other causes, such as lack of oxygen. It is therefore an end result, not a cause of death. Apoptosis is very different, because a positive mechanism is triggered that results in contraction of the nuclear chromatin, and very commonly the degradation of chromosomal DNA to nucleosomes (see Tomei & Cope 1991; Wyllie 1992). It is now clear that apoptosis, or a similar process, regularly occurs in development when there is a need to eliminate cells. This happens, for instance, between the digits of the hand and foot, during the innervation of muscles, and in the elimination of lymphocytes during the development of the mature immune system.

A variety of cell-damaging agents can induce apoptosis. Therefore it is a possibility that endogenous or intrinsic damage from any of the causes outlined in Chapters 3 and 4 might also result in apoptosis. If this is so, the apoptosis could be regarded as a positive beneficial process that gets rid of cells before they can do damage to other cells. There are many reasons why a cell on the pathway to cell death from some other cause could have deleterious effects on surrounding cells. After all, cells in tissues act in concert, so one that is not responding to stimuli, or producing aberrant molecules that affect surroundings cells, could remove itself by a suicide mechanism. This would be beneficial, especially if that cell can be replaced. Similarly, apoptosis could be a defence against tumour formation, if it was to remove a cell that had progressed some way towards malignancy.

There is at present much work being done on apoptosis in the context of cancer research, development of the immune system, the nervous system or in other locations. Although there has been discussion of apoptosis in the context of cell and tissue ageing (e.g. Monti et al. 1992), there appears to be no direct evidence that ageing is associated with apoptosis. One exception may be the regression of the size of the thymus in adulthood. In other situations, occasional apoptotic cells may not have been detected amongst an excess of functional cells. The ageing of human cells *in vitro* is not associated with apoptosis, unless that occurs after cells detach from their substrate. An important question that needs to be answered is whether intrinsic events, such as accumulation of altered proteins, damage to DNA, abnormal mitochondria and so on, which would eventually cause a cell to become senescent, instead precipitate more rapid death by apoptosis.

There is one interesting context where a suicide mechanism may be very important. Of the large number of oocytes that are produced in the female mammal, most regress by a process known as atresia. It has been proposed that apoptosis is involved in ovarian atresia (Tilly et al. 1991; Hurwitz & Adashi 1992). There are good reasons to believe that quality controls may exist in germ line cells, and that imperfect cells may be removed by a suicide mechanism (see Chapter 6).

6

Genetic programmes for ageing

In earlier chapters the fundamental differences between germ line and somatic cells have been stressed. The germ line that provides continuity from generation to generation does not age, except in unusual circumstances. At least a proportion of the cells must be potentially immortal, with this immortal cell lineage stretching back to the progenitors of existing species. Whatever provides the basis for the indefinite proliferation of germ cells is therefore shut off in most, and perhaps all, somatic cells. In other words, regulatory changes controlled by the genetic programme and development come into play. Since different mammalian species have very different lifespans, the nature or extent of these genetically controlled changes must vary from species to species. This is in contrast to the germ line, which must have common species independent features keeping it in a permanently juvenile state.

THE PRESERVATION OF THE GERM LINE

Germ line cells retain their juvenile state irrespective of parental age, with a few notable exceptions that are discussed later. In effect, a steady state must be maintained in the sense that the fertilised egg of one generation has the same potential for development as that of any earlier or later generation. One might expect genetic or other abnormalities to gradually accumulate from generation to generation, and this has given rise to the question, 'Why are babies born young?' (Bernstein 1981). In answer to this question, it has been proposed that a major function of meiosis is to provide the opportunity for the repair of DNA damage, and that this is an essential mechanism for avoidance of ageing in the germ line (Bernstein & Bernstein 1991). Although there may be truth in this, it cannot be a complete explanation. DNA damage arising immediately before or during meiosis could be repaired, but other DNA damage either is not heritable, and so would not be transmitted to the meiotic cell, or is indeed heritable, and so would not be recognised as abnormal. This is simply due to the fact that once a mutation is fixed in DNA in a germ cell lineage, there is no mechanism by which it could be recognised as different from the previously existing DNA sequence. We know, of course, from the study of genetics of innumerable organisms that new mutations are passively transmitted through meiosis. Thus, we can conclude that new mutations that do not drastically affect fertilisation and development will often survive, although they are, of course, subject to selection pressures at every stage.

The power of selection is likely to be crucial in keeping germ cells in a juvenile or steady state. Such selection may act on cells with defective genes or chromosomes, and also those with abnormal cytoplasm. The possibility of escalating errors in RNA and proteins was discussed in Chapter 4. As well as this, there could be defects in mitochondria or other organelles, or in the cell membrane. It is also possible that there are quality control mechanisms that in some way screen germ cells for the existence of defects or errors and eliminate those that are imperfect. Programmed cell death in development and other contexts occurs by the suicide mechanism known as apoptosis (see Chapter 5). This can be triggered by particular signals; for example, apoptosis can be induced in lymphocytes by glucocorticoids. Similarly, there may be mechanisms in germ cells that can detect abnormalities in RNA and proteins (arising from DNA mutations, or transcriptional or translational errors) and can trigger an elimination or suicide mechanism. It is even possible that there are special proteins that induce cell death, only if they contain one of several errors (see Note 6.1). By such editing or proof-reading devices, only those germ cells that have normal cytoplasmic components would be transmitted to the next generation. In the ovaries of young female mammals there are very large numbers of oocytes, far more than could ever develop to mature eggs. It is well known that most of these oocytes regress by a process known as atresia, and it is possible that this is the manifestation of a strong selection mechanism that eliminates all oocytes that are not normal. It is less clear whether a similar process occurs in the male germ line. Certainly many abnormal spermatozoa are produced that are non-mobile and could never fertilise an egg.

The mechanism of transmission of mitochondrial DNA (mtDNA) from generation to generation is an unsolved problem. It is known that the mutation rate in mtDNA is about 10 times higher than that in chromosomal DNA, and also that there are a large number of mtDNA genomes per cell. Mitochondrial DNA from different individuals often has distinct sequences (recognised as restriction enzyme site polymorphisms), yet the mtDNA in one individual is homogeneous. This suggests that there is some special mechanism in the female germ cell lineage, possibly in the oocyte, that in some way ensures homogeneity in mtDNA sequence. It is possible that one functional mtDNA sequence is amplified and all others eliminated. This could be a mechanism to rid oocytes and eggs of defective mtDNA genes, which would otherwise proliferate in the developing organism, with severe effects on its respiration.

Epigenetic events are known to be reversed in the germ line, probably at or just prior to meiosis. The inactive X in female germ line cells is reactivated at that time. Also, genomic imprinting (see Note 6.2) from the previous generation is reversed by the time new gametes are produced. There are also dramatic changes in DNA methylation during spermatogenesis and early embryogenesis, the significance of which is not yet understood (Monk, Boubelik & Lehnert 1987; Razin & Cedar 1993). It is possible, however, that meiosis plays a positive role in the detection of defects in methylation. In particular, it has been suggested that the methylation in important regulatory regions might provide a signal for the initiation of recombination. The formation of hybrid DNA in this region, and the activity of a DNA maintenance methylase (see Chapter 3) would then repair an epigenetic defect in one chromosome (in this case, loss of methylation), provided its homologue was normal (Holliday 1984b, 1987, 1988b). This could be another mechanism for ensuring that the DNA in the fertilised egg is normally programmed for subsequent development.

AGEING OF THE GERM LINE?

In simple animals the monitoring of germ line cells may be less effective, possibly because the distinction between germ line and somatic cells is less strict. In classical experiments on ageing carried out many years ago, Lansing (1947) studied the lifespan of the offspring of rotifers from mothers of different age. He found that if eggs were continually selected from elderly mothers, then the maximum lifespan declined in each generation. This strongly indicated that deleterious changes are occurring in the germ line of this organism. (Note, however, that in a natural environment, most offspring are produced from young mothers; see Chapters 1 and 7.) Attempts have been made to repeat Lansing's experiment using other species, such as *Drosophila*. The results have been somewhat controversial, but a consensus view is that the Lansing effect is only likely to be seen in very simple invertebrates such as rotifers (for discussion, see Bell 1988; Finch 1990). In long-term experiments with *Drosophila,* offspring from late-breeding flies were sequentially selected, and the result was a gradual increase in longevity (see Kirkwood & Rose 1991; Rose 1991; Partridge & Barton 1993). This is the opposite of the Lansing effect, and it depends on the selection of new combinations of genes (from stocks that were originally heterozygous) that increase longevity.

An influence of parental age on the longevity of offspring has not been reported in mouse or man. There are, however, other age-related parental effects. The best known is the elevated frequency of Down's syndrome amongst offspring of human females of increasing age (Hassold & Jacobs 1984; Hassold & Chin 1985). Most human embryos with chromosome abnormalities do not survive, but trisomy 21 of Down's syndrome, as well as sex chromosome abnormalities, are notable exceptions. The female oocyte remains in the dictyotene stage of meiosis from about seven months gestation to the time the oocyte matures, completes meiosis and forms a haploid egg. At the dictyotene stage crossing-over between homologous chromosomes has already occurred, and such crossovers are important to ensure normal chromosome disjunction when the first nuclear division of meiosis takes place. It is probable that during the extensive period between seven months gestation and the late period of female fertility (~40 years), some crossovers terminalise and are lost. This would greatly increase the likelihood of chromosome non-disjunction, especially in a small chromosome with few crossovers. Trisomy 21 can therefore be regarded as an example of an age-related error in the germ cell lineage, which is not eliminated.

In human males, there are also age-related genetic defects. It is known that some dominant mutations are transmitted significantly more frequently by elderly fathers (Risch et al. 1987). The frequency of some of these mutations increases more than linearly with age. This suggests that the mutation frequently does not relate directly to the number of cumulative cell divisions in the germ cell line, but instead is caused by some other influence, such as a reduction of efficient DNA repair in germ line cells of males of increasing age.

REGULATED CHANGES IN SOMATIC CELLS

In Chapter 3 various maintenance mechanisms were described. Since these mechanisms are genetically specified, it follows that somatic cells of short-lived species should have less efficient maintenance than long-lived ones. If the assumption is made that the germ cells of all species have very effective maintenance mechanisms, then there would be greater down-regulation of some of these in the somatic cells of short-lived species than those of long-lived ones. This is a prediction that could eventually be tested by experiment. In comparative experiments that have so far been done (see Chapter 7), it has been shown that there is a direct relationship between the efficiency of DNA repair (and several other maintenance processes) and species lifespan.

Reduction in the maintenance efficiency of somatic cells may not be a sudden event. It may well be that embryonic cells, like germ cells, maintain themselves very effectively, but that as differentiation proceeds, certain enzymes or other proteins required for maintenance are reduced or fully repressed. Mouse embryonal stem cells, like germ line cells, are reported to be immortal, whereas other embryonic cells have finite lifespan (McLaren 1992). The disappearance of telomerase activity may be a good example of loss of a maintenance activity. Following this event, DNA telomeric sequences are lost at each somatic cell division, and this may be the reason for their limited lifespan. Established mouse fibroblast cell lines have a very weak DNA excision-repair mechanism, but there is evidence that primary cell cultures from embryos are capable of excision repair, and that this ability is lost after further growth (Ben-Ishai & Peleg 1975; Peleg, Raz & Ben-Ishai 1977). Although most cultured somatic cells have finite growth, it is possible that some somatic cells, such as stem line cells, can divide indefinitely *in vivo* (see Chapter 5). If such cells do exist, they must retain a much more effective means of maintaining their viability than fully differentiated cells. It is possible that the transition of potentially immortal ('uncommitted') cells (see Chapter 5 and Note 5.3) to cells with finite growth potential is due to the loss of specific maintenance functions, including DNA telomerase activity.

THE CONCEPT OF PROGRAMMED AGEING

In the gerontological literature, as well as in many popular articles on ageing, it is often stated that there must be a programme that determines lifespan. This programme is thought to be specifically controlled; for example, there could be a molecular clock that in some way measures chronological time, or measures the number of cumulative cell divisions. The possible molecular basis for such a specific programme has rarely been discussed; nor has it been explained why such a programme should have evolved. It has been proposed that ageing is part of the developmental process, so that development to a fertile adult is subsequently followed by a decline in reproductive capacity, and then senescence (see, for example, Kanungo 1975). According to this view there are genes that control early development, and other late-acting genes that control the running down of the developmental programme.

The menopause in human females is cited as evidence of a programme that turns off the reproductive system some years before signs of ageing become manifest. Menopause is, of course, genetically programmed, but it appears to be unique to higher primates. It may be a secondary adaptation to ensure that the successful upbringing of slowly developing offspring is not jeopardised by the continued breeding of elderly mothers (see Chapter 7). The general concept of the running down of a programme for development is more a description of what is seen rather than any specific explanation. It does not explain why fully developed reproducing adults should not have the potential to breed indefinitely, nor why the late-acting genes should have evolved in the first place (see also Chapter 7).

It is possible to envisage molecular clocks that count cell divisions. It has been suggested that these may be important in the genesis of the population of a particular size, or in the temporal determination of sequential events in development, which are tied to division (Holliday & Pugh 1975). One such developmental clock depends on the sequential methylation of repeated DNA sequences, in such a way that one repeat is modified at each division. Only at the end of a given number of divisions, specified by the number of repeats, is the regulatory signal induced (see Note 5.4). There are no compelling reasons to believe that such clocks operate in controlling the ageing process, although it is easy to see how the timing of a clock could be altered during evolution by unequal genetic exchange, which could generate more or fewer repeats (Holliday 1990).

From studies on the longevity of inbred mice, which are genetically homogeneous, we know that lifespans have a considerable range (see Fig. 1.2), even though the specific cause of death is not the same for all animals. This by itself demonstrates that there cannot be an accurate clock or strict programme. The results indicate that the mechanisms that are responsible for the onset of senescence and the time of death are likely to be heterogeneous, or multiple, and that the time course for their onset is variable. It is very likely that stochastic events are in part responsible for determining the onset of specific changes. In the majority of organisms ageing is a fairly gradual process extending over a significant proportion of the total lifespan. The activities of genes in controlling longevity are likely to be diverse, rather than specific, since they affect different cell types and organ systems in such a way that a general synchrony is seen in the decline of one relative to the other.

In some specific cases, ageing occurs much more rapidly over a relatively short period of time. This suggests that there is a much more significant regulation of the process, and it is then more appropriate to apply the term 'programmed ageing'. In the next section examples of such rapid programmed ageing are described, but it is important to realise that these represent a small number of species that have adopted a specific life style. In particular, we find that certain organisms that are semelparous (i.e. reproduce only once) subsequently enter a well-defined process of senescence, which rapidly leads to death.

PROGRAMMED DEATH FOLLOWING REPRODUCTION

The breeding behaviour of the Pacific salmon provides the best-known example in a vertebrate of programmed senescence and death (reviewed by Finch 1990). The young hatch in small freshwater streams and migrate down to the sea. The mature adults return to the same stream to spawn. Huge amounts of eggs and sperm are produced, and subsequently the fish deteriorate very rapidly. This process, triggered by the hormonal changes associated with breeding, leads to the rapid disintegration of the tissues of the salmon, which is very different from the normal course of ageing. This situation probably evolved because the fish can produce significantly more offspring than they would if they retained their ability to return to the sea, with the possibility of renewed breeding the following year. It is also possible that the decay of the dead parental salmon provides organic material that greatly augments the food chain in the streams and rivers, which will ultimately benefit their offspring. The programmed death of the Pacific salmon is likely to be a recently evolved secondary adaptation, because other salmon that travel much shorter distances to spawn, such as the Atlantic salmon, may return to the sea from their breeding grounds and can subsequently breed again.

Another example of programmed ageing in a vertebrate is provided by several species of *Antechinus,* which are small marsupial mice (see Diamond 1982). In this case it is only the male who succumbs. After an intense and exhausting period of breeding, in which as many females as possible are fertilised, the male does not recover and dies at the age of about a year, which is much earlier than other marsupial species of similar size. The adaptation that is beneficial is to put maximum resources into the first cycle of breeding, rather than into several successive cycles. In small mammals that have many predators, offspring are born mainly to young adults, so the life cycle of male *Antechinus* is just a more extreme case of this.

A third example of death after breeding is provided by *Octopus hummelincki* (Wodinsky 1977). After laying her eggs, the female ceases to feed and dies a few weeks afterwards. This is a programmed event tied to the breeding cycle, because if the optic glands are removed after spawning, females resume feeding and growth and have a greatly extended lifespan. The secretions of the optic glands are responsible for the induction of programmed ageing. It is possible that the life style of this species of octopus is an adaptation partly related to putting maximum reserves into egg production, but also to provide an appropriate ecological niche for surviving offspring. Individual octopi tend to feed in a well-defined area of their general habitat, so the removal of the mother may provide living space and food that would otherwise be unavailable. This is, in fact, the argument used by Weismann to explain the evolution of ageing in all species. What we see in the real world is that only in special cases does the death of parents directly benefit their offspring.

Whereas breeding followed by death is the exception in vertebrates or advanced invertebrates, it is common in insects and some other invertebrates (see Finch 1990). Annual plants also demonstrate a similar breeding life style, whereas perennials do not. Many plants propagate themselves clonally, that is, by vegetative reproduction, but it is unusual to find trees that do so. One example is the banyan tree, which can establish a colony of clonally-related individuals, expanding from the centre, whereas most other trees grow to form a single structure that continually increases in size. These other trees eventually die because their weight becomes too great for their mechanical strength (see Note 6.3). Death of large trees releases a new habitat, that is, part of the forest floor, in which offspring can compete. The widespread absence of clonal propagation in trees may provide, in effect, further examples of ageing that support Weismann's viewpoint, since if the trees simply consolidated their living space and were not removed, there would be much less opportunity for their offspring to survive, and for generation to follow generation.

GENE MUTATIONS THAT CHANGE LIFESPAN

Mutations often affect the viability of organisms and thereby reduce their lifespans. Most of these mutations have nothing to do with ageing per se because a defect in an essential biochemical pathway is very likely to jeopardise the organism's survival. This is well illustrated by many of the several thousand known inherited defects in humans (McKusick 1990). Unfortunately, many of them have a severe effect on one or another normal bodily function, and therefore significantly reduce expectation of life. Subsets of these mutations, however, do have distinct phenotypic effects that may relate to natural ageing. A few can be regarded as premature ageing syndromes, that is, they appear to accelerate some of the normal processes of ageing. The pleiotropic effects of the mutations that cause Werner's syndrome and progeria (Hutchinson–Gilford syndrome) are remarkable, and the identification of the genes involved may well provide important insights into the understanding of ageing in normal individuals. These inherited conditions are described and discussed in Chapter 8.

It is amongst animals that are used for the experimental study of ageing, such as *Drosophila* species and the nematode *Caenorhabditis elegans,* that one can expect new mutations to be identified that have defined effects on ageing and lifespan. In fact, it is commonly believed that the identification of such 'gerontogenes' is the most likely route to a better understanding of ageing. This optimism may be misplaced, since ageing is due to changes in many cell and tissue systems, and it is unlikely for a single gene or a small number of genes to control all these events. Nevertheless, at least one gene mutation, *age*-1, has been identified in *C. elegans* that has the effect of increasing lifespan (Johnson 1987, 1990a,b; Friedman & Johnson 1988). Initially it seemed likely that this mutation also reduced fertility; thus, it was possible that the resources normally used for reproduction were channelled instead into maintenance of the soma. However, it was subsequently shown that two genes are involved in determining the phenotype: one affects fertility, and the other (*age*-1) affects lifespan. The problem with the study of animals in the laboratory is that mutants such as *age*-1 may have reduced fitness, because they lack phenotypic characteristics very necessary for survival in a natural environment. For example, they may lack the ability to deal with toxic chemicals in food or in their surroundings, or to escape predators. In the constant environment provided by the laboratory, such mutants may survive very well; in fact, their survival may benefit from the loss of 'dispensable' functions, which may nevertheless be indispensable in a natural environment. However, this does not detract from the importance of identifying single-gene mutations that affect lifespan.

In general, inbred organisms have shorter lifespans than outbred ones. It would therefore be expected that genetic heterozygosity can provide the necessary variability for the selection of altered lifespan. This has been achieved in *Drosophila* in experiments based on the reasonable assumption that flies that breed later in life would also have an increased lifespan. By selecting eggs from late-breeding females, it was possible to increase the lifespan by 29% after 15 generations of selection (Rose 1984, 1991). It is to be expected that these long-lived flies have slower development and certainly begin to breed later than the founder population. In a natural environment this would almost certainly be disadvantageous, since most offspring are produced from young animals. The selection experiments in the laboratory may well favour flies in which resources normally used for rapid development and breeding are instead diverted to later reproduction and better maintenance of the soma.

An important mouse model has been developed in a series of studies in Japan since 1970 (reviewed by Takeda, Hosokawa & Higuchi 1991): the senescence-accelerated mouse (SAM). This is an inbred strain derived from AKR/J mice, but the genetic basis of the observed phenotype is not clear. These animals have normal development and do not exhibit strong premature senescence, but rather show a more rapid development of the senescent phenotype. These changes include senile amyloidosis, degenerative joint disease, senile osteoporosis, cataracts, a decline in immune response, and deficient learning and memory. The major biochemical changes that have so far been detected are increases in lipid peroxidation in serum, liver and skin. There is also an increase in chromosome abnormalities in comparison to normal animals. Further studies of SAM mice are likely to reveal much important new information, but it is equally important to identify the genetic basis for the changes seen.

MICROBIAL MODELS

Genetic and molecular manipulation is easiest in microbial systems, so it is not surprising that a number of interesting mutations relevant to ageing have been uncovered and studied in detail. One of the most significant examples directly relates to the hypothesis for mammalian cellular ageing. Normal yeast strains maintain their chromosomal telomeres, but through a study of the mechanisms whereby this was achieved, it was possible to construct strains that were defective in telomere maintenance. These cells could divide a limited number of times, but then became 'senescent' and died out (Lundblad & Szostak 1989). If the criterion for cellular ageing is a period of normal cell division followed by cessation of growth, then these yeast cells provide an excellent model for what might be happening in diploid mammalian cells. Normal yeast cells have also been used for studies on cellular ageing. Yeast cells produce daughters by budding, so the question arises as to whether or not the mother cell can produce buds indefinitely. In experiments where the daughter cells were sequentially removed, it was clear that the mother cell produces only a limited number of buds, and this is not due to the accumulation of bud scars. The yeast cell eventually 'ages' after a given number of buds have been produced (commonly in the range 20–40), and this has been exploited as an experimental system in which the activity of genes in cells of different ages can be studied (Egilmez, Chen & Jazwinski 1989). Unlike the telomerase model, the yeast budding system seems somewhat far removed from mammalian cell ageing, not least because dividing cells in higher organisms do not have a mother–daughter relationship as do yeast cells.

In earlier studies it has been shown that wild-type vegetative cells of the fungus *Podospora anserina* have a defined lifespan (Marcou 1961). Like higher organisms, this fungal species survives only by continuous sexual reproduction, which can only occur in non-senescent cells. It was shown that the ageing of *Podospora* was due to cytoplasmic changes, and recent molecular studies demonstrate that the senescent cells have abnormal mitochondrial DNA (see Note 6.4). Since the organism is an obligate aerobe, normal mitochondria are essential for growth. The events that give rise to abnormal DNA are not clear, but when such mitochondria do appear they spread through the coenocytic fungal mycelium, eventually displacing the normal population. In this regard, the senescence seen is almost certainly not a good model for the ageing of populations of mammalian cells *in vitro* or *in vivo*, since mitochondria cannot move from cell to cell. Another related fungus, *Neurospora crassa*, can grow indefinitely without sexual reproduction, but certain mutants are known that will grow in culture only for a defined length of time. These are called *nd* (natural death; Sheng 1951) and *leu*-5, which is temperature sensitive (Printz & Gross 1967). The latter is known to produce abnormal proteins, especially at high temperature (35°), so it is possible that its limited growth at this temperature is due to protein error feedback. Evidence that this might be the case was obtained by the rapid appearance of temperature-sensitive glutamic dehydrogenase and inactive cross-reactive material during the time the cultures were slowing down and dying (Lewis & Holliday 1970). The *nd* mutation is recessive and can be maintained in a heterokaryon. Monokaryotic *nd* cultures grow at a constant rate for a while and then cease growth, just as does wild-type *Podospora*. However, in the case of *nd*, evidence was obtained that the fidelity of protein synthesis was reduced in senescent mycelia (see Note 6.5).

Certain ciliates, such as *Paramecium aurelia*, grow by binary cell fission, but in the absence of sexual reproduction clones of cells eventually degenerate and die. These organisms are unusual in having a micronucleus (analogous to the germ line) and a macronucleus (analogous to the soma). Sexual reproduction or autogamy (self-fertilisation) depends on meiosis in the micronuclei, nuclear fusion and the elimination of superfluous micronuclei and the macronucleus. New macronuclei are generated in both processes, and the cells become rejuvenated. The senescence seen in long-lived clones of *Paramecium* is in part due to irreversible changes in the macronucleus, although cytoplasmic abnormalities are also present. The organism provides interesting experimental material for the further study of cellular ageing, and existing knowledge has been fully reviewed by Bell (1988). The ciliates are very unusual in that their cells are highly complex. In multicellular organisms functions are divided between different cell types, whereas in *Paramecium* and other ciliates, complex phenotypic features, such as movement, feeding, reproduction and so on, are all carried out by a single highly specialised cell with a unique nuclear organisation. The characteristics of the cells of ciliates are so completely different from that of the individual cells of multicellular differentiated eukaryotes that *Paramecium* may not be an appropriate model for the study of cellular ageing in higher organisms.

GENES AND AGEING

In this chapter various features of genetic effects on ageing have been briefly discussed, and two major conclusions emerge. First, ageing in mammals and many other higher organisms is such a complex process, involving many cells, tissues and organs, that many genes are likely to be involved in determining the whole aged or senescent phenotype. Many of these genes relate in one way or another to the maintenance mechanisms discussed in Chapter 3, and also in Chapter 4. Mutations in these genes may have important effects on the normal biochemistry or physiology of the organisms, and although viability and lifespan may be strongly affected, there need be no obvious effect on ageing per se. Second, there is a somewhat heterogeneous collection of examples, where either a fairly specific genetic programme for ageing exists, or where individual mutations appear to strongly influence ageing and lifespan. The examples of programmed ageing are the exception rather than the rule, unless of course we use 'programme' in the very general sense of 'genetic determination'. Mutations that affect ageing are also likely to be a heterogeneous collection. Some may increase ageing at the expense of some other important phenotypic trait that is not obvious in the artificial laboratory environment. Others may have fairly trivial physiological effects; for example, one that reduced the efficiency of food utilisation might have analogous effects to calorie deprivation and thereby increase lifespan (see Chapter 7). Others, perhaps more interesting, may have specific effects on cellular processes likely to be important in ageing, for example, the proteolytic degradation of abnormal proteins, the defence against free radicals, or DNA repair. What does seem clear is that the view that there are a few critical 'gerontogenes' that somehow control ageing is somewhat naïve, and it is unlikely that laborious and time-consuming experiments to identify such genes will yield much insight into the molecular and cellular changes that are responsible for ageing.

Mutations that increase lifespan will always excite public interest, but from a scientific perspective, those that accelerate ageing may be equally or more important in future studies of molecular changes during ageing. Another area of research that will become very important in the future is the investigation of differences between germ line and somatic cells. It was argued that important maintenance or quality control mechanisms exist in germ line cells, but at least some of these are switched off in somatic cells. The study of the genetic regulatory mechanisms involved is of crucial importance.

Finally, there are a number of microbial models, which illustrate how lifespan can be manipulated by genetic methods, and their underlying biochemistry and molecular mechanisms can be successfully investigated. Nevertheless, the study of ageing in microbial species is not the same as the study of fundamental genetic and biochemical processes that are common to all cells. The pattern of ageing relates strongly on the life style of the organism, and what is seen in a simple organism may be completely different from the complex series of changes that occur in the ageing of birds or mammals.

7

The evolution of longevity

In discussion of the origins and existence of animals that have a soma of finite lifetime, there are two interrelated evolutionary problems. The first is the evolution of ageing itself. As we saw in Chapter 1, this must have occurred very early on in evolutionary history in simple invertebrates. It is also likely that mortal somatic cells and tissues appeared in more than one invertebrate taxonomic group. If this is the case, then any discussion of the evolution of ageing in vertebrates is to some extent misleading, because they were derived from animals that already had a mortal soma. Therefore, discussion of the 'evolution of ageing' in higher organisms is really discussion of the evolution of longevity. The second problem is the modulation of longevity. What determines the length of life of mammalian and other species? We know in many cases that evolution resulted in an increase in longevity, and it would be surprising if it did not sometimes favour a reduction in lifespan. The evolutionary forces moulding longevity are obviously complex, but a major factor is the fecundity of the organism in relation to the environment, or ecological niche, that the organism inhabits.

EARLY THEORIES OF THE EVOLUTION OF AGEING

The evolution of ageing was first seriously discussed by August Weismann (see Kirkwood & Cremer 1982), who proposed that ageing was a necessary adaptation to ensure the further evolution of offspring by natural selection. He argued that in the absence of ageing, the available environmental resources would be dominated by parents, or their parents, leaving insufficient opportunity for offspring to survive to adulthood and to reproduce. Darwinian natural selection depends on the existence of novel variation amongst offspring, and this can only occur if generation follows generation. Therefore any species consisting of individuals that could survive indefinitely would be at a dead end, and could not evolve to occupy new or changing environments. According to Weismann's views, such species would become extinct, and only those that had the positive adaptation of an ageing soma would have the potential to survive and evolve further.

At first sight, the argument is very persuasive, and for a long time Weismann's views were not seriously challenged. As we saw in Chapter 6, there are some organisms in which the ageing of the parents is of positive benefit to the offspring, but these are only a minority of the innumerable species in which ageing is seen. The fundamental point that Weismann missed is, in fact, implicit in the Darwinian theory itself. Natural selection can only operate if more offspring are produced than can survive to adulthood and reproduce, and selection depends on the greater chance of survival of specific genetic variants within a whole population. Darwin knew from countless examples that the mortality rate amongst offspring was usually very high, so the *proportion* of offspring that themselves survive to reproduce was often quite small. Therefore under normal circumstances few individuals would survive indefinitely, and there would be no need to evolve the adaptation of ageing.

This fundamental flaw in Weismann's argument was made clear by Medawar (1952) more than 70 years later. He pointed out that in a natural environment, small animals and birds have very high mortality, and aged individuals are rarely, if ever, seen. Ageing of these organisms becomes obvious only when they are kept in the artificial environment of a zoo, or a laboratory cage, where they are regularly fed, protected from predators and so on. Medawar therefore realised that a quite different explanation for the evolution of ageing was required. He was dismissive of Weismann's ideas, but in formulating his alternative theory he made an assumption that genes can act early or late in adulthood. He correctly stated that most offspring are born to young parents, and very few to old parents, simply because very few individuals survive to be old. He therefore argued that a deleterious mutation that acted late in life would not be subject to natural selection, because such mutations are not transmitted, or very rarely transmitted, to offspring. He therefore concluded that the process of ageing was due to the accretion over long periods of time of many late-acting deleterious mutations, or to put it more graphically, senescence was merely the rubbish bin for all such mutations. The accumulation of mutations of the type he envisaged is simply dependent on time, and it is not easy to see how a large number of mutations could be present in individual lineages. In other situations, mutations that reduce fitness are eliminated, and mutations that increase fitness are selected and spread through the population. Therefore, 'late-acting' mutations are effectively neutral. An important issue is the likelihood of accumulation of sufficient neutral mutations with the required late-acting phenotype.

Williams (1957) was unaware of Medawar's publication when he formulated his own theory for the evolution of ageing a few years later. He realised that if late-acting mutations were to accumulate, there must be some selective force involved. He therefore proposed that there is an important class of *pleiotropic* mutations, which have beneficial effects on young animals but harmful effects on old or older animals. Such mutations would be selected because most offspring are born to young parents, and their accumulation would result in senescence and death. Williams's argument was later shown by Hamilton (1966) to have a sound quantitative basis in terms of population genetics and evolution.

The Medawar–Williams–Hamilton theory is based on the supposition that mutations can be early acting or late acting. It is important to note that this can only apply to organisms that *already have a finite lifespan*. It is not applicable to organisms that develop to an adult and that can maintain themselves indefinitely; in other words, organisms that do not age. Simple animals such as flatworms (planarians) have a pool of totipotent stem cells, which can replace all fully differentiated cells (see Chapter 1). Their remarkable powers of rejuvenation means that they are in a non-ageing steady state (Child 1915; Sonneborn 1930). In such animals early-acting and late-acting mutations probably have no meaning, but one could envisage mutations that abolished the normal function of the stem cells, and therefore produced an animal with finite lifespan (in the same way as irradiation, which kills the stem cells, produces an apparently normal animal with finite survival time [Lange 1968]). However, it is clear from the discussion of early- and late-acting genes that Medawar, Williams and Hamilton were all primarily concerned with ageing in higher organisms. Therefore, they were really discussing the *modulation of ageing,* or the evolution of longevity. It is essential to understand that the concepts of early- and late-acting mutations are meaningful only in the context of an organism that has a finite span of life; in other words, in an organism that already ages (Kirkwood & Holliday 1979). In view of this, it is surprising that the Medawar–Williams–Hamilton theory continues to be accepted in more recent discussions of the 'evolution of ageing' (Charlesworth 1980; Rose 1991; Partridge & Barton 1993).

THE DISPOSABLE SOMA THEORY

In Chapter 1 the alternative strategies for the survival of multicellular species were discussed. On the one hand, an organism can develop to adulthood and keep reproducing as long as it survives in a hazardous natural environment, and in a protected environment it has the potential to survive indefinitely. On the other hand, the organism develops to adulthood and reproduces, but does not have the potential to maintain its soma indefinitely. In a protected environment signs of senescence appear after a given time, and the animal dies.

The concept of a 'disposable soma' originally developed from a molecular theory about the stability of protein synthesis (Kirkwood 1977; Kirkwood & Holliday 1979). When life first evolved there was a very major problem in maintaining molecular integrity. Initially macromolecules would not have been made with great accuracy, but these error-containing molecules must in some way have been able to perpetuate themselves, and also to evolve to become more accurate. The accuracy we see in present-day organisms depends on highly specific enzyme systems, and also various proof-reading devices that would recognise errors and correct them (see Chapter 3). DNA is replicated with very high fidelity, and as well as efficient proof-reading there are also many enzyme-based repair mechanisms. We also know that the synthesis of RNA and proteins is less accurate by several orders of magnitude. With regard to ageing, the question at issue is the stability of the process of protein synthesis. As was explained in Chapter 4, error-containing molecules involved in RNA and protein synthesis can generate more errors, and if this error feedback is above a certain level, then the proportion of defective molecules continually increases to a lethal error catastrophe. This is a problem early organisms had to solve during the evolution of increasing accuracy. Have present-day organisms evolved to a state in which error feedback is negligible, and the likelihood of an error catastrophe is vanishingly small or nonexistent? Alternatively, have they evolved to a point where protein synthesis is stable, but there is a reasonable likelihood that this stability will be lost through error feedback? Discussion of these possibilities has generated much sophisticated theoretical analysis (see Note 4.4), but a reasonable conclusion is that organisms evolved an optimal level of accuracy in protein synthesis, compatible with continued survival, but not too dependent on elaborate energy-consuming proof-reading or correction devices. From this, one can also argue that the optimum may well vary between organisms, depending on their different survival requirements. It could well be that cells are in a metastable state in which the accuracy is maintained, but there is a given probability that errors will increase to a level that kills the individual cell. Such cell death could contribute to or possibly be a major component of ageing.

Initially, the disposable soma theory took into account accuracy in macromolecular synthesis, the energy investment required to obtain a given level of accuracy and the trade-off between such investment and the production of offspring. Thus, rapidly-breeding, short-lived animals made a smaller investment in accuracy than slow-breeding, long-lived ones. The metabolic cost of repair of macromolecules was an obvious inclusion into the theory (Kirkwood 1981), and later on many other types of mechanism were discussed in terms of the maintenance of the adult organism, which were reviewed in Chapter 3. Today, the disposable soma theory includes the considerable metabolic expense of all such maintenance mechanisms, and the trade-off between this expense and the investment of resources into growth to adulthood and reproduction. On the basis of reasonable assumptions (Kirkwood & Holliday 1986b; Kirkwood & Rose 1991) it can be shown that to achieve maximum Darwinian fitness the optimum investment in maintenance results in finite survival. The investment required for infinite survival, that is, an immortal phenotype, will always reduce fitness (Fig. 7.1).

It can be argued that the disposable soma theory for the evolution of ageing is merely a special case of Williams's pleiotropic mutation theory. The argument is based on the supposition that any mutation that reduces maintenance and increases fertility, or vice versa, is pleiotropic and subject to natural selec-

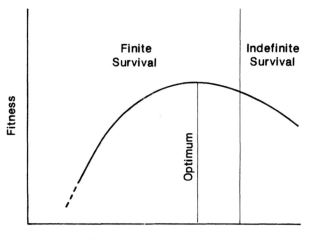

Figure 7.1. The relationship between investment of metabolic resources and Darwinian fitness of the organism. The resources necessary for indefinite survival, without ageing, will always reduce fitness below the optimum value (see Kirkwood & Holliday 1986b; Kirkwood & Rose 1991; Rose 1991).

In a formal sense this may be so, but in practice single mutations are unlikely to have such diverse effects. For example, any mutation that increases the efficiency of the immune system is hardly likely to reduce fecundity directly, and one that improves the gestation success rate will in all probability have no effect on the immune response. The fact is that maintenance and reproduction, taken in the broadest sense, encompass so much of the organism's metabolism that selection for a change in longevity is much more likely to be based on the interactions between the innumerable genes that determine any major feature of the phenotype. This in turn ultimately depends on the evolution of individual genes, but these need not have any pleiotropic effects.

One of the major predictions of the disposable soma theory is that the immortal germ line cells should have more elaborate cellular maintenance mechanisms than somatic cells. They might, for example, have significantly higher accuracy in DNA, RNA and protein synthesis, greater powers of detection and repair of faulty macromolecules, more efficient recognition and degradation of aberrant proteins and so on. We do know that cells undergoing meiosis are highly specialised, and there is a very high frequency of recombination between homologous chromosomes. It has been suggested that this is essential for the rejuvenation or prevention of ageing of germ cell lineages (see Chapter 6). It may well be that meiocytes have additional specialised functions, relating perhaps to protein metabolism, the preservation of normal mitochondria and so on. At present, there is no information about the possible difference between accuracy of RNA and protein synthesis in somatic and germ line cells. There is much information about mutation frequencies in the germ line and a few studies using somatic cells, but in no case is there a direct comparison using the same gene and organism. There is very little known about the comparative efficiencies of DNA repair in the two types of cell.

Recently, however, it has been shown that somatic cells progressively lose telomeric DNA at the ends of the chromosome arms (Harley et al. 1990, and see Chapter 4). This is thought to be due to the absence of the enzyme telomerase. Telomeric DNA must be maintained by telomerase in germ line cells. Thus, in this case, a prediction of the disposable soma theory seems to be confirmed, since somatic cells dispense with a maintenance mechanism necessary for continuous survival. Another example of loss of maintenance in somatic cells is the decline in the level of 5-methyl cytosine in DNA (see next section).

A second prediction of the disposable soma theory is that long-lived organisms will have more efficient somatic maintenance mechanisms than short-lived ones. A third prediction is that, within major taxonomic groups, there should be a direct relationship between the maximum lifespan of an organism and its reproductive potential, or fecundity. Much evidence in support of these second and third predictions is now available from mammalian species, and this is reviewed in the following two sections.

COMPARATIVE STUDIES OF THE EFFICIENCY OF MAINTENANCE

Several major investigations have been carried out on metabolic processes that form part of maintenance, using cells from mammalian species with different lifespans. At least six separate studies have dealt with DNA repair. In the first, Hart & Setlow (1974) established primary fibroblast cultures from seven mammalian species. These were irradiated under constant conditions with ultraviolet light, and the extent of DNA synthesis outside the S phase (unscheduled DNA synthesis [UDS], see Note 7.1) was determined, which depends on the measurement of ^{3}H-thymidine incorporation into DNA using autoradiography. The results shown in Figure 7.2 clearly show a relationship between longevity and extent of repair. This study was followed up by the study of DNA repair in primates (Hart & Daniel 1980; Hall et al. 1984). The fibroblasts of six species, and the lymphocytes of eight, demonstrated a clear correlation between the extent of repair of UV damage and longevity. The two rodent species *Mus musculus* and *Peromyscus leucopus* are similar in size, but have maximum lifespans of about three and eight years, respectively (Sacher & Hart 1978). It was also shown that repair synthesis after UV treatment was significantly greater in *Peromyscus* (Hart, Sacher & Hoskins 1979). Another study by Kato et al. (1980) examined a larger number of mammalian species, also using primary cultured fibroblasts. In these experiments the authors concluded that there was no obvious relationship between

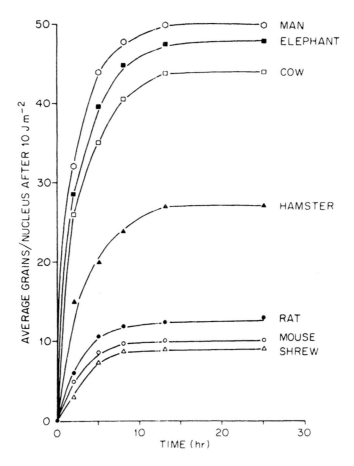

Figure 7.2. The amount of repair synthesis as a function of time after a dose of 10 J/m² UV light for seven different mammalian species. The cells used were primary cultures of fibroblasts. (Reproduced with permission from Hart & Setlow 1974.)

repair synthesis and the lifespan of a donor. Scrutiny of this publication, however, illustrates several of the problems of comparative studies. It is known, for instance, that fibroblasts from different parts of the body are not identical; for example, they vary in their *in vitro* lifespan. Also, the age of a donor animal may be an important factor in DNA metabolism. Another variable is the number of cell generations that have occurred *in vitro* before the experiment was done. In the study by Kato et al. (1980) cells were taken from different parts of the various animals, and neither the ages of the donor nor the ages of the cells *in vitro* is specified. Also, it has been pointed out by Tice & Setlow (1985) that if Chiroptera and Primates are removed from the data, there is a reasonable correlation between UDS and lifespan.

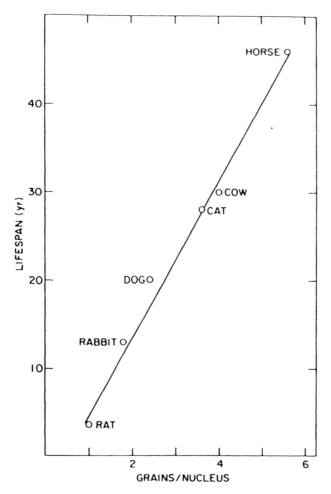

Figure 7.3. The amount of DNA repair synthesis after exposure to UV light in primary lens epithelial cells, and donor lifespan. (Redrawn from Treton & Courtois 1982.)

A much better controlled study is of particular significance, because the cells used are normally exposed to UV light during the animal's lifetime (Treton & Courtois 1982). These cells were from the lens epithelium, established as primary cultures. As in the first study, there is a close relationship between DNA repair synthesis and the maximum lifespan of the donor (see Fig. 7.3).

In another study a different method was used to estimate repair, which depended on the bromodeoxyuridine photolysis technique to measure both the number of repaired lesions and the patch size (Francis, Lee & Reagan 1981). Amongst 20 species, a reasonably good correlation between extent of repair and longevity was demonstrated. In conclusion, it can be stated with some certainty that there is a good correlation between the ability to repair DNA damage and longevity. Out of six studies this relationship is demonstrated in five, and the other is equivocal. A more detailed discussion of nearly all these data is available in Tice & Setlow (1985).

Several other investigations on DNA metabolism in different species have also been carried out. Carcinogenic hydrocarbons are modified by detoxifying enzymes, and in this metabolic process intermediates are formed that are subsequently degraded. It is well established that these intermediates can be highly reactive oxidised molecules, such as epoxides, which can bind to and react with DNA to produce chemically altered bases. Clearly in this detoxification process it is important to minimise the amount and half-life of the reactive intermediates. When 7-12-dimethyl benzanthracene (DMBA) was incubated with fibroblasts from several mammalian species, different amounts of the active intermediates bound to DNA were detected (Schwartz & Moore 1977). The shortest-lived species (rat, followed by guinea pig and rabbit) had the greatest amount of bound carcinogen, and in the longest-lived (human and elephant) there was almost no detectable binding. Intermediate activity was seen using cow fibroblasts. This suggests that the detoxification process is more efficient in the longer-lived species, presumably because there is very little accumulation of reactive intermediates. Another study documented the mutagenicity of activated DMBA (Schwartz 1975). This assay measured the ability of fibroblasts to convert DMBA to a form that was mutagenic to V79 target cells. (HPRT⁻ mutants resistant to 8-azaguanine were screened.) The number of mutations seen was inversely correlated with the lifespan of the fibroblast donor species. It is well known that spontaneous neoplastic transformation of rodent cells occurs much more frequently *in vivo* and *in vitro* than in human cells. Also, innumerable carcinogens have been shown to transform mouse, rat or hamster cells, but not to transform human cells (see Holliday 1987). The resistance of cells to neoplastic transformation is, of course, a vital part of cell maintenance (see Chapter 3).

The enzyme poly-ADP ribose polymerase is known to be an important component of DNA metabolism, although its exact role is still a matter for debate. It has recently been shown that the extent of poly ADP ribosylation by cell-free extracts from different species is greatest in longer-lived species and least for shorter-lived ones (Grube & Burkle 1992). These studies were based on 13 mammalian species, and there were no exceptions to the direct relationship between enzyme activity and maximum lifespan.

The level of 5-methyl cytosine (5-mC) residues in the DNA of mammalian species is fairly similar, but when primary diploid cells were cultured *in vitro,* this level was not maintained (Wilson & Jones 1983). In the mouse cell populations there is a very rapid decline in total 5-mC, and many experiments have shown that these cells have become senescent after about 10–15 population doublings. The decline in hamster cell populations is much slower, and these cells grow to 25–40 population doublings. Human cell populations maintain their 5-mC more successfully since the level declines much more slowly during serial passaging, and these cells usually achieve 50–70 population doublings (see also Fairweather et al. 1987). In this example, maintenance correlates better with the *in vitro* lifespan of fibroblasts than with the donor lifespan, since the lifespan of hamsters is similar to that of mice. Measurements of the total 5-mC content in the DNA of cells from donors of different age show a very slow decline in humans, and a significantly faster decline in mouse tissue (Wilson et al. 1987).

The inactivation of one X chromosome in female mammals is associated with the methylation of CpG islands of housekeeping genes. There is evidence that the inactive X in mice is reactivated with fairly high frequency during ageing, whereas reactivation of the human X is a much rarer event (see Chapter 3). Presumably the reactivated genes on the mouse X chromosome have lost 5-mCs in island regions, although this has not yet been examined directly. If so, the results imply that epigenetic controls based on DNA methylation are much more stably maintained in man than in mouse (Holliday 1989a, and see Note 4.9).

Other observations have been concerned with cell lifespan. Mammalian erythrocytes have no nucleus, and it is well known that they have a finite lifetime in the blood. Their lifespan is presumably determined by cytoplasmic proteins, and particularly the proteins making up the membrane. As a membrane deteriorates, soluble components start to leak out. These membrane changes are recognised by the liver, and old erythrocytes are removed from the blood. It has been shown that the lifespan of erythrocytes in blood is directly related to the maximum lifespan of the species (Rohme 1981), so presumably the membrane proteins or other features of the cytoplasm are much more stably maintained in long-lived species. In the same study, the longevity of dividing fibroblasts in culture was determined. (These results are shown in Fig. 5.4) Although there is much controversy about the mechanisms of fibroblast senescence, there is agreement that fibroblast ageing is related to the ageing of the organism (see Chapter 5).

The cross-linking of collagen is one of the best biomarkers of ageing. Chemical analysis of the number of mature cross-links in collagen in bovine and human skin shows a similar biophasic increase in the two species, but the *rate* of cross-linking is very significantly higher in bovine than in human skin (Fig. 7.4). There is about a three-fold difference in bovine and human lifespans, whereas the figure shows a much greater difference than this in the rate of cross-linking. The rate of growth of cattle is very high in comparison to human growth, and it may well be that this provides an explanation for the faster bovine cross-linking. It is well known that cross-linking of collagen in rat tail tendon continually increases over the three-year lifespan (see Fig. 2.1), and one would expect the rate of cross-linking to be very much greater than in bovine or human collagen, although exact comparisons using identical

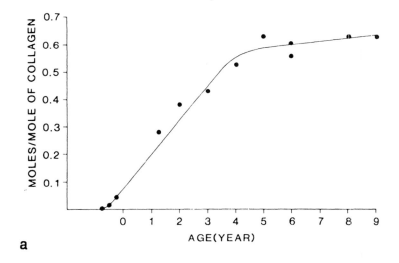

a

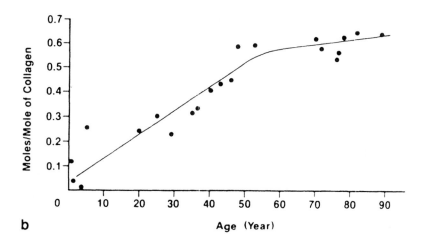

b

Figure 7.4. The formation with age of mature collagen cross-links in (a) bovine and (b) human skin. The cross-linked structure was identified as histidino-hydroxylysinonorleucine (HHL). (Reproduced with permission from Yamauchi, Woodley & Mechanic 1988.)

procedures have not yet been done. Nonetheless, the experiments indicate that the integrity of collagen structure must be much better maintained in long-lived species, and the same could be said of other proteins such as crystallin in the lens, elastin in arteries and so on. We do not yet understand how this better maintenance is achieved, but it is unlikely to be a function of amino acid sequence, and the secondary and tertiary structures of collagen in different species are very similar. It is more likely that small metabolites or oxygen free radicals interact with collagen and other long-lived proteins to produce cross-linking, and that the extent of this interaction varies between species. Also, the extent of non-enzymic glycation of proteins over a given interval of time may well be higher in short-lived mammals than in long-lived ones (see Chapter 4).

It has often been claimed that maximum lifespan is some function of metabolic rate. In general terms this may well be true, because large mammals have a low metabolic rate and usually have a long lifespan, and small mammals have a high rate and short lifespans. Recently, the claim has been made much more specific by comparing oxygen free radical damage to DNA with the metabolic rate. This was done by measuring the amounts of a derivative of thymine that is produced by free radical attack, excised from DNA and excreted in the urine (Adelman et al. 1988). The four species examined were mouse, rat, monkey and man, and the correlation held good. Since the amount of oxygen consumed per kilogram of body weight would also be related to the quantity of free radicals generated, it was concluded that free radicals are likely to be an important cause of ageing. However, it is dangerous to draw such a specific conclusion when three variables are correlated. In this case the variables are (1) the amount of DNA damage/kilogram of body weight, (2) metabolic rate and (3) lifespan. In the four species examined, all are correlated with each other, so it is possible that in other species only two of the three variables are correlated. This is likely because important exceptions occur in the supposed correlation between longevity and metabolic rate. Bats are small mammals that generate large amounts of energy in their flight muscles, and therefore have a very high metabolic rate, although it is much lower when at rest. Several small bat species kept in zoos, or marked and released in the wild, show that they can have lifespans as long as 20 years, which is several times greater than the lifespan of most small rodents (see also 'Fecundity and longevity', below). Thus, in this case the correlation breaks down because metabolic rate per se relates to the life style of the organism, rather than having any direct relationship to longevity, or to maintenance. A very important addition to the data would be to analyse the amount of free radical damage in a long-lived small mammalian species, such as a bat. The disposable soma theory of ageing would predict that the amount of free radical damage to DNA in this animal would relate to its lifespan, rather than to the metabolic rate of the organism. It may well be that metabolic rate is a measure of the *production* of free radicals, but the actual damage suffered depends on the defence mechanisms available (see Chapter 3), and this in turn may relate to the efficiency of maintenance and to longevity.

In earlier studies Cutler (1984) examined the concentration of β-carotene and retinol in brain tissue and serum of six primate and seven other mammalian species. These substances were chosen because they have antioxidant and possibly anticancer activity. A correlation between the concentration of carotene and longevity was observed, but not of retinol. More recently, evidence has accumulated in the literature that retinol has a very important role in morphogenetic signalling during the control of gene expression. This casts doubt on its possible importance in defence against oxygen free radicals. In another study, Cutler (1985) found that there was an inverse relationship between the peroxide-producing potential of brain or kidney tissues and the longevity of 24 mammalian species. This was done by aerobically inhibiting tissue homogenates and measuring their rate of auto-oxidation, which is related to the amount of peroxidisable substrate. The results indicate that effective defence against oxygen free radicals is related to longevity.

The comparative studies that have so far been published on various aspects of metabolism relating to maintenance are listed in Table 7.1. In all except one case, the efficiencies seen correlate with the lifespan of the species. The correlation also holds for the survival in time of erythrocytes in the blood, and the *in vitro* lifespan of fibroblasts.

FECUNDITY AND LONGEVITY

The third major prediction of the disposable soma theory is that there is a strong inverse relationship between maximum lifespan and fecundity, that is, the potential number of offspring produced during the maximum period of reproduction. This follows from the trade-off argument, namely, that there are limits to the metabolic resources available to any animal, and these may be preferentially channelled into preservation of the soma for a long period, or alternatively channelled into rapid growth to adulthood and rapid production of offspring. In general terms, it is easy to see that such a relationship exists: mice, which start to breed at about 6 weeks old, can produce many litters in a year, and their lifespan is about 3 years; domestic cats start breeding at about 1 year, produce two or three litters annually and have a lifespan of 15–20 years; herbivores commonly have one offspring a year and live 30–40 years, and so on. Nevertheless, there have been no serious attempts so far to examine the relationship in more detail, which may in part be due to the difficulty of obtaining exact information. Estimating the longevity of mammals is fraught with difficulties. In their natural environment most individuals rarely achieve their maximum lifespan, whereas they do so in a protected environment. Therefore captive zoo animals provide the best data, and the most up-to-date and reliable records have been compiled by M. L. Jones (1982, and personal communication; see also Note 7.2). Nevertheless, there are still problems in estimating longevity. For most species only a few individuals have lived out their natural lifespans in zoos. This is true, for instance, of the great apes, where recorded or estimated lifespans are 48–59 years. One could ask what the lifespan of a very small number of humans would be if kept under the same conditions. A reasonable estimate might be 70–80 years; how-

Table 7.1. *Studies of cells from different mammalian species*

Parameter measured	Cell type or source	No. spp.	Result[a]	Reference
Longevity *in vitro*	Fibroblasts	8	+	Rohme (1981)
Longevity *in vivo*	Erythrocytes	11	+	Rohme (1981)
DNA repair after UV light	Fibroblasts	7	+	Hart & Setlow (1974)
DNA repair after UV light	Fibroblasts	34	0[b]	Kato et al. (1980)
DNA repair after UV light	Fibroblasts and lymphocytes	6	+	Hall et al. (1984); Hart & Daniel (1980)
DNA repair after UV light	Fibroblasts	20	+[b]	Francis et al. (1981)
DNA repair after UV light	Lens epithelium	6	+	Treton & Courtois (1982)
Mutagenicity of activated DMBA	Fibroblasts	6	–	Schwartz (1975)
Binding of activated DMBA to DNA	Fibroblasts	6	–	Schwartz & Moore (1977)
Poly-ADP ribose polymerase	Lymphocytes	13	+	Grube & Burkle (1992)
γ-ray–induced ADP ribosyl transferase	Lymphocytes	12	+	Pero, Holmgren & Persson (1985)
DNA methylation decline	Fibroblasts	3	–	Wilson & Jones (1983)
Auto-oxidation and peroxide formation	Homogenates of brain and kidney	20 9	– –	Cutler (1985) Cutler (1985)
Superoxide dismutase	Liver, brain and heart	14	0[c]	Tolmasoff et al. (1980)
Concentration of carotenoids	Serum	13	+	Cutler (1984)
Metabolic rate and oxidised DNA bases	Urine	4	–	Adelman et al. (1988)

[a]The parameter measured is in each case compared with the maximum lifespan of the species. + indicates positive correlation; – indicates negative correlation; 0 indicates no obvious correlation.

[b]In the study by Kato et al. (1980) a correlation exists if species of the orders Primates and Chiroptera are excluded. Also, the correlation seen by Francis et al. (1981) is stronger if some primates are excluded. (See Tice & Setlow [1985] for a full discussion; see also this text.)

[c]There is no correlation between longevity and specific activity of superoxide dismutase (SOD), but the level of SOD in relation to metabolic rate does correlate with lifespan.

ever, from among the millions of records of human lifespans, we know the maximum is 120 years. Not only is the sample size enormously greater, but there may also be rare genetic variants with increased longevities. We have the same problem with domesticated animals, as there are very long recorded lifespans of cats, dogs, rabbits and so on. Again, the sample size is enormous, and there is genetic variability. In the case of experimental animals, it is usual to determine the lifespan of cohorts of, say, 50–200 animals. Is the maximum lifespan the highest achieved by any individual or the average lifespan of the surviving 5 or 10% of individuals? In Figure 7.5 the values used are primarily based on the zoo records, which in turn are based on small sample sizes.

There are also difficulties in estimating the reproductive potential of each species. The important parameters are gestation period, litter size, interlitter interval, the time to reach sexual maturity and the period of fertility. In many cases lactation delays subsequent pregnancy, but if the offspring die young, then females become pregnant much sooner. In Figure 7.5 it is assumed that the interlitter interval is constant, that is, survival of the offspring is longer than the period of lactation. There is also the problem of declining fertility with age. In human females the menopause is the end of fertility, but in most species there is simply a gradual decline of fertility with age. Records for most species are not available, so a rule-of-thumb assumption is made, namely, that females breed for two-thirds of their lifespan after they have become sexually mature. This is undoubtedly an oversimplification, because mammals that breed rapidly as young adults probably lose fertility proportionally earlier in their lifespans than more slowly breeding ones, but it is very hard to quantitate such differences.

In spite of the various difficulties in estimating the longevities and reproductive potential, Figure 7.5 shows that there is a very clear relationship between longevity and fecundity in eutherian mammals. This relationship holds despite the fact that the 47 genera included in the figure have very different anatomies and biological life styles. The overall trend seen confirms a major prediction of the disposable soma theory.

There are, however, some important exceptions to the inverse relationship between longevity and fecundity. Amongst the rapidly-breeding rodents, there are some that live two to three years and others with similar fecundity that live significantly longer. (The longevity difference between mice of similar size, *Mus* and *Peromyscus,* has provided the basis of several important experimental studies.) Rabbits, squirrels and bears (nos. 7, 12, and 15 in Fig. 7.5) seem to produce more offspring than would be expected from their respective longevities. It is worth noting, however, that although zoo specimens of the grey squirrel have lived for 23 years (Jones 1982) their survival in the wild is surprisingly brief (see Fig. 1.3). Animals that have relatively low fecundity in relation to longevity are some bats, some herbivores and

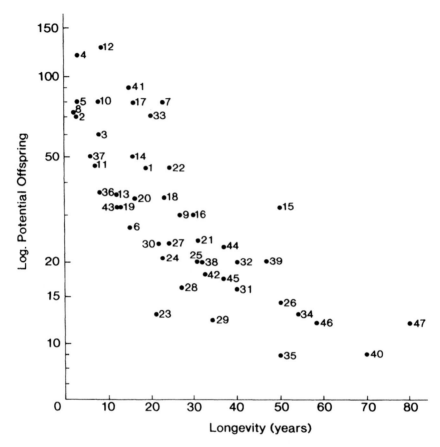

Figure 7.5. The reproductive potential and longevity of 47 genera of eutherian mammals. Data obtained primarily from *Asdell's Patterns of Mammalian Reproduction* (Hayssen, van Tienhoven & van Tienhoven 1993), *Walker's Mammals of the World* (5th Ed., Nowak 1991) and M. L. Jones (1982, and personal communication). Reproductive potential is the maximum number of offspring that might be produced under ideal conditions and assumes there is no infant mortality (see text). Longevity is based primarily on the maximum lifespans of limited numbers of individuals kept in captivity (see text). Both estimates are approximate. The genera are numbered as follows: 1, chinchilla (*Lagidium*); 2, hamster (*Cricetus*); 3, white-footed mouse (*Peromyscus*); 4, laboratory mouse (*Mus*); 5, laboratory rat (*Rattus*); 6, beaver (*Castor*); 7, grey squirrel (*Sciurus*); 8, lemming (*Lemmus*); 9, porcupine (*Hystrix*); 10, guinea pig (*Cavia*); 11, hare (*Lepus*); 12, rabbit (*Oryctolagus*); 13, fox (*Vulpes*); 14, dog (*Canis*); 15, bear (*Ursus*); 16, lion (*Panthera*); 17, cat (*Felis*); 18, otter (*Lutra*); 19, skunk (*Mephitis*); 20, badger (*Meles*); 21, *Hyaena;* 22, raccoon (*Procyon*); 23, fruit bat (*Eidolon*); 24, collared fruit bat (*Rousettus*); 25, flying fox (*Pteropus*); 26, camel (*Camelus*); 27, vicuña (*Vicugna*); 28, deer (*Cervus*); 29, giraffe (*Giraffa*); 30, sheep (*Ovis*); 31, ox (*Bos*) or *Bison;* 32, horse (*Equus*); 33, Pig (*Sus*); 34, *Hippopotamus;* 35, *Rhinoceros;* 36, elephant shrew (*Elephantulus*); 37, hedge-

some primates. This can probably be related to their evolved life style. Small ground-living mammals are frequently killed by predators, but the flight of bats and their habit of hanging from the roof of caves allows them to escape predation. Reproduction, however, is more difficult, and many species produce only one offspring per year. The evolution of this life style has clearly been associated with a considerable increase in longevity. It is sometimes asserted that the longevity of bats is due to winter hibernation, but tropical species do not hibernate and are equally long lived. In the case of herbivores, there is usually a single offspring that is very well developed at birth and very soon able to escape from predators. Also, they develop rapidly and breed sooner than other mammals of comparable size. The evolution of primate longevities and reproduction are considered in a later section ('Primate evolution').

ENVIRONMENTAL EFFECTS ON AGEING

It is very well established in mice and rats that a restricted diet results in a significant extension of their lifespan (reviewed by Schneider & Reed 1985; Holehan & Merry 1986). Animals fed with only 50–60% of an unrestricted, or ad libitum, diet live 40–50% longer than those fed ad libitum. Also, many of the pathologies associated with ageing are very significantly delayed, or occur less frequently. It is also established that food restriction severely reduces fecundity in these animals. In any natural rodent environment, food resources are likely to fluctuate greatly, so that animals are sometimes able to take advantage of available food and reproduce, whereas at other times, when food is scarce, reproduction slows down or does not occur at all. It is well known that a food glut allows small animals such as rodents to reproduce rapidly and produce enormous populations. When food finally runs out, the animals became stressed and infertile, or they may be compelled to migrate.

From an evolutionary standpoint, the effects of nutrition on longevity make biological sense. Individuals during a food glut will put all their resources into reproduction; but if the environment changes to one where food is very scarce, reproduction is suspended, because it would be biologically disadvantageous to attempt to rear offspring under such conditions. In conditions of

Caption to Figure 7.5 (*cont.*)

hog (*Erinaceus*); 38, sloth (*Choloepus*); 39, Minke whale (*Balaenoptera*); 40, elephant (*Elephas*); 41, mouse lemur (*Microcebus*); 42, brown lemur (*Lemur*); 43, marmoset (*Callithrix*); 44, macaque (*Macaca*); 45, baboon (*Papio*); 46, orangutan (*Pongo*); 47, man (*Homo*). The list includes 10 rodents, 10 herbivores, 10 carnivores, 7 primates, 3 chiroptera and representatives from other orders. Species were chosen that provided the most reliable data, and usually there is only one representative of closely related species (large cats, great apes, sheep and goats etc.). For tabulated data, see Holliday (1994).

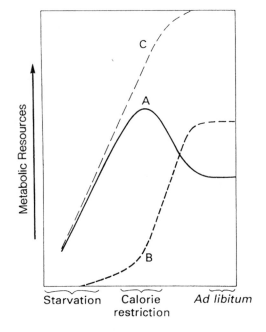

Figure 7.6. A suggested relationship in rodents between food intake and metabolic investment
in reproduction (B), and all processes required for maintaining
the adult body or soma (A).
Curve C is the sum of A and B,
and this is assumed to be a constant proportion of the total energy available to the animal.

semi-starvation, the best strategy for the animal is to survive by investing
what energy resources are available into increased maintenance. This will tide
the animal over the lean period, and the reappearance of adequate food will
then allow it to reproduce. It is known from experiment that rats maintained
on a low-calorie diet, but subsequently fed ad libitum, can reproduce at significantly greater ages than control animals fed ad libitum throughout. It thus
seems reasonable to suppose that life extension induced by calorie deprivation
is an adaptive mechanism that in most natural environments will result in the
optimum production of offspring (Holliday 1989b; Turturro & Hart 1991).
The channelling of available food resources into reproduction and maintenance is illustrated in diagrammatic form in Figure 7.6. The nature of the regulatory processes that could switch animals from investment in reproduction
to investment in maintenance, and vice versa, are at present only a matter
for speculation. A number of experimental studies of calorie-restricted and
unrestricted rats or mice have been undertaken. These include investigation
of spontaneous DNA damage (Randerath et al. 1991), the maintenance of genetic fidelity (Hart et al. 1990), the frequency of somatic mutation (Dempsey,
Pfeiffer & Morley 1993), the activity of hepatic cytochrome P450 enzymes
(Alterman et al. 1993), and the proportion of heat-labile or oxidised proteins
(Goto, Ishigami & Takahashi 1990; Youngman, Park & Ames 1992). These
and many other studies from R. W. Hart's group cannot be reviewed in detail
here. There is not much doubt that they will provide information and insight

into the nature of the ageing differences in calorie-restricted animals and those fed ad libitum. So far, all the indications are that natural ageing is slowed down in the calorie-deprived animals (Turturro & Hart 1991).

Recently, another strong environmental effect on ageing has been uncovered. In this study opossums (*Didelphis*) living in the relatively restricted environment found on islands were compared with the same species living in the more open and variable mainland environment (Austad 1993). The main finding was that the island variety, or subspecies, was larger and had greater longevity than the mainland population. The latter were more at risk from predators, and generally encountered more hazardous conditions than the former. This would also be predicted by the disposable soma theory, because under potentially hazardous conditions the best strategy is to reproduce fast and invest fewer resources into maintenance of the soma. On the other hand, the island population survived better by increasing body size and investing greater resources into maintaining the soma for a longer period. The results obtained in this study support the prediction of lifespan extension in island populations, but there were no data available that documented reproductive rates of the animals in the two different environments.

PRIMATE EVOLUTION

In the evolution of primates, there was an increase both in size and also in longevity. However, there does not seem to be a simple relationship between longevity and fecundity. Some small primates, such as the mouse lemur (*Microcebus*) and the marmoset (*Callithrix*), have fairly short lifespans and breed fairly rapidly. There may be two or three offspring per litter and more than one litter per year. Other small primates such as lorises (e.g. *Nycticebus* and *Loris*) and the squirrel monkey (*Saimiri*) produce one offspring a year, and also have relatively short lifespans. It is possible that the adaption of an arboreal habitat, the evolution of intelligence and memory, and the effective protection of single young offspring by the mother have all resulted in a considerable decline in mortality. In other words, the probability of an infant surviving to adulthood and itself reproducing would be much higher than in many other mammals of comparable size and longevity.

The fecundity of the great apes has hardly changed over evolution. With an interbirth interval of two to three years in the chimpanzee, gorilla, orangutan and human female (Hayssen et al. 1993), the maximum number of offspring per female is very similar. In the case of primates, it seems that longevity relates much more strongly to the time taken to reach sexual maturity, that is, the rate of growth to adulthood. This is illustrated in Figure 7.7 for 13 representative species. It therefore seems that the increase in longevity is more strongly related to a declining rate of development, rather than to a reduction in fecundity. A strong argument can be made that human evolution occurred by a process known as 'neoteny' (de Beer 1958). This is a process whereby

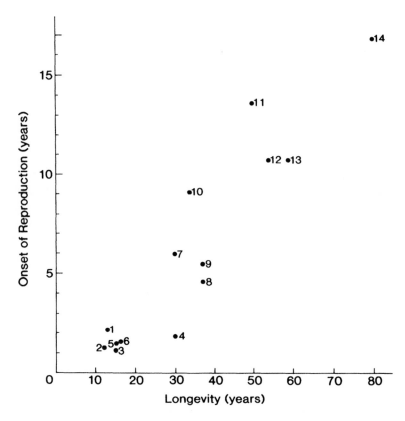

Figure 7.7. The relationship between the age of onset of reproduction of female primates, including the period from fertilisation to birth, and the longevity of the species. 1, slow loris (*Nycticebus*); 2, slender loris (*Loris*); 3, mouse lemur (*Microcebus*); 4, brown lemur (*Lemur*); 5, squirrel monkey (*Saimiri*); 6, marmoset (*Callithrix*); 7, spider monkey (*Ateles*), 8, macaque (*Macaca*); 9, baboon (*Papio*); 10, gibbon (*Hylobates*); 11, chimpanzee (*Pan*); 12, gorilla (*Gorilla*); 13, orangutan (*Pongo*); 14, human (*Homo*). See legend to Figure 7.5 for sources of data.

development is slowed down so that the adult of a newly evolved species has characteristics of a young individual of an earlier species. On anatomical grounds, humans more closely resemble young apes than adults, but of course, sexual reproduction must be brought forward relative to overall body development. In the evolution of primates as a whole, there is clearly a slowing down of development, an increase in the time to reach sexual maturity and an increase in longevity. These in turn probably relate to the evolution of learning ability and intelligent behaviour, which better adapt the species to a hostile environment. This latter adaptation is also associated with a decline in mortality and a lower rate of reproduction. There is also a strong argument made by Wilson (1991) that rates of evolution correlate with brain size in vertebrates. He proposed that the more an animal relates its behaviour to the environment in which it finds itself, the faster it will adapt in Darwinian terms to this environment. Intelligent awareness of the environment would be correlated with brain size, and therefore an increase in brain size would be related to the rate of evolutionary change. This could apply strongly to primates and help to explain the rapid evolution of brain size. Furthermore, the evolution of language as an extension of behaviour would further accelerate successful adaptation to the environment. The acquisition of language and other skills is favoured if development to adulthood slowed down. Thus we would expect puberty to occur later in humans than in the great apes, and this in turn relates to delayed reproduction and a lengthened lifespan.

Whether this change in longevity evolved rapidly is a matter for debate. A claim has been made that there was a particularly rapid increase in lifespan during the evolution of humankind, which leads to the conclusion that rather few genes may be involved in determining lifespan (Cutler 1975). This claim and conclusion demand critical examination. In particular, the estimation of maximum lifespans of the larger primates is difficult, as there are so few specimens kept in zoos. As was explained previously, it is likely that zoo lifespans of chimpanzees and gorillas (about 50 years) correspond to human lifespans of 70–80 years, not 120 years, which is a frequently quoted figure (see Note 7.3). Thus, instead of at least a two-fold difference in longevity, there is more likely to be a difference of 20–30 years. This weakens the view that an increase in longevity evolved rapidly. The question of the number of genes involved in lifespan determination is discussed in Chapter 9.

After humans evolved, it is generally agreed that they adopted a hunter-gatherer life style. They probably lived in small groups of interrelated individuals, just as do some of the great apes today. It is interesting to consider the demography of such human populations in relation to reproduction and longevity. The early aboriginal population of Australia comprised a reasonably homogeneous population that broke up into many subgroups or tribes. Archeological evidence suggests that the aborigines arrived about fifty thousand years ago, and at the time the white settlers arrived in 1788, it is estimated that the total population was between three and five hundred thousand. If we assume that the founder population was no more than a few hundred individuals, say a thousand, and that there are on average four or five generations per century, then it can be easily calculated that the average increase in population size per generation is 1.0028 or 0.28% (see Note 7.4). This is scarcely different from a steady state population. To maintain a population in a steady state, each reproducing female must herself produce, on average, two reproducing adults. On the basis of certain simplifying assumptions (see Note 7.5), it can be calculated that the average expectation of life might be only 25–30 years for the individuals who survive infancy and early childhood. If infant and childhood mortality is taken into account, then this figure would be very significantly lower.

Environments in which hunter-gatherers live will fluctuate between those in which food is relatively plentiful and much harsher conditions in which food and water are very scarce. These changes will have very strong influences on reproduction and life expectancy. When human societies evolved agriculture and effective division of labour, the environment became much more constant and more favourable. The contrast between such communities and the hunter-gatherers of Australia is well demonstrated by Polynesian communities. Land on many of the Pacific islands is very fertile, and there is always a plentiful supply of fish. As a result the populations are not in a steady state. The expectation of life increases, as does the average number of offspring produced per female; hence the population begins to rise exponentially. When living space is restricted, this population growth cannot be sustained, resulting in social stress, war and colonisation of other islands. The social pressures of population increase are probably the driving force for all human migrations.

The evolution of human longevity would be the result of an interplay between many phenotypic characteristics, including the ability to obtain a food supply from the environment, which depends in turn on intelligent behaviour; the extent of mortality in infancy and childhood; the likelihood of adults producing sufficient offspring to maintain the population, and the social organisation of hunter-gatherer groups. In developed nations today the maximum lifespan is about 25–30 years greater than the expectation of life. In early hunter-gatherer communities the difference would have been very much greater, perhaps as much as 60–70 years. Nevertheless, in such communities, some adults would survive to old age. In small communities of very mixed age, kin selection and the evolution of altruistic behaviour is likely to become very important. To ensure survival of offspring, it would be counterproductive for females to keep reproducing, rather than successfully rearing existing children. The evolution of the menopause was in all likelihood an adaptation to ensure survival of younger individuals in the society (see Note 7.6). It is probable that non-reproducing females cared not only for their own children, but also for grandchildren or other young relatives, just as they often do today. The existence of the menopause indicates that even in primitive hunter-gatherer societies, some individuals achieved a longevity which was a considerable proportion of the maximum lifespan.

AGEING IS AN ADAPTATION

It is sometimes stated that Weismann's theory of the evolution of ageing is adaptive, and that other theories such as the Medawar–Williams–Hamilton theory and the disposable soma theory are non-adaptive. Strictly speaking, this distinction is artificial, because if we consider the wider question of the strategy for survival, then the transition from an immortal to a mortal soma could be either disadvantageous or adaptive. This relates specifically to the organism (Chapter 2). Strong powers of repair, replacement and regeneration, which can in principle produce a non-ageing organism in a steady state, were replaced by the evolution of specialised features of the soma, which gradually became incompatible with replacement, repair and regeneration. It was a better survival strategy to evolve a design that allowed the organism to exploit the environment and produce offspring, rather than to try to survive intact for long periods, or indefinitely. Once the mortal soma had appeared in diverse taxonomic groups, then of course natural selection could modulate the balance between reproduction and somatic maintenance. In different contexts, longevity increased during evolution, or it may have decreased.

In simple terms, the disposable soma theory is simply one device to increase fitness and to favour the transmission of selfish genes. Ageing itself is not directly subject to selection, because as we have seen, in natural environments almost all individuals are produced by young parents and very few individuals reach old age. Nevertheless, the soma has to survive long enough to transmit its genes, so there are strong selective forces that optimise reproduction in terms of survival of the soma, that is, the organism's biological lifespan.

The comparative studies that have so far been done strongly indicate that cells from long-lived species have greater powers of maintenance than those from short-lived ones. Of course many more cellular and molecular studies are required to test the general proposition in greater detail, and in principle many of these can be done now using modern experimental procedures. The effects of nutrition and the environment on longevity are also compatible with, and indeed support, the disposable soma theory. These experiments demonstrate the importance of studies on ageing in the wider contexts of population ecology and demography. The whole theory of natural selection is based on the fact that more offspring are produced than can survive in any given environment, and the potential number of offspring that can be produced is related to the lifespan of the organism. It is therefore quite surprising that Darwin had little to say about ageing and lifespans when formulating his theory of natural selection.

8

Human disease and ageing

Experimental gerontology is the study of the ageing of animals or plants, or of their cells in culture. In the many reviews and books discussing this field, very little, if any, attention is paid to the many human pathological changes that occur during ageing. Yet this pathology comprises a vast resource of material that is very relevant to any understanding of ageing. Since the serious initiation of anatomical medicine in the 16th and 17th centuries, information has been collected about pathological changes in the human body, some of which occur predominantly in old individuals. With the introduction of the use of the microscope and the realisation that organs and tissues are made up of cells, the science of histology made possible much more detailed descriptions of pathological changes. In more recent times, the electron microscope and biochemical analysis provided yet more information, and to this must now be added the extensive molecular studies that are being applied in more and more contexts.

For the most part this material is documented in specialised journals that deal with different organ or tissue systems. The total information is vast, and certainly enormously greater than the whole gerontological literature. Much may not be relevant to an understanding of ageing, but because so many pathological changes are age-related, a considerable proportion is not only relevant, but also provides very important insight into the cell, tissue and body changes that occur in human ageing, and into mammalian ageing in general. This chapter certainly cannot attempt any serious review of all this material. The intention is to highlight particular pathological conditions that appear to relate to the failure of maintenance, that is, to demonstrate that the anatomy and histology of many tissues and organs are incompatible with an indefinite survival. For the most part, the summarised information is not documented with specific references, since it can be found in standard monographs on pathology. Extensive references to the pathological changes during the ageing of both humans and rats are listed in the *Handbook of Physiology in Aging* (Masoro 1981), and the recent *Oxford Textbook of Geriatric Medicine* (Grimley Evans & Franklin Williams 1992) is an invaluable source of information.

THE BRAIN, VISION AND HEARING

Brain tissue consists predominantly of neurones and neuroglia (glial cells). Neurones are post-mitotic cells, which are unable to divide in response to damage, or to replace abnormal or dead neurones. The total number of neurones in the brain gradually declines with age. There are several types of neuroglia, and these are capable of division. Their detailed functions are unknown, but they appear to have an active role in supporting neurones. In damaged brain, the neuroglial cells known as astrocytes act somewhat like fibroblasts in the rest of the body and can form glial scars. However, no collagen or equivalent extracellular protein is produced, so the scar consists entirely of cellular material. Glial cells cultured *in vitro* have finite division potential (see Chapter 5). Other neuroglia known as oligodendrocytes produce myelin and appear to have the function of maintaining the normal myelin sheaths of the neurones in the central nervous system.

An essential feature of the brain is the extreme localisation of function. When combined with the very limited capacity for regeneration and repair, this means that the brain is a sensitive target for pathological changes, including those that occur during ageing. Some of the commonest causes of brain damage are associated with defects in the vascular supply. Although the human brain is about 2% of the total body weight, it uses about 20% of the total oxygen supply. Any serious decline in blood supply results in hypoxia, and in the space of a few minutes irreversible brain damage will occur. Excess blood pressure (hypertension) may result in rupture of a major blood vessel, and the resulting haemorrhage is followed by a failure of the oxygen supply to vital parts of the brain. Abnormalities of the vascular system with resulting brain damage (that is, one or more strokes) are a major cause of death in old age.

The ever-increasing proportion of elderly individuals has focused attention on the importance of Alzheimer's disease and related brain disorders. The prevalence of Alzheimer's disease increases steadily from late middle age, with about 5% of individuals over the age of 65 affected and about 20% of individuals over the age of 80 years. There is much confusion about the diagnosis of the disease in relation to age. It is conventional to diagnose Alzheimer's disease in relatively young individuals whose brains lose normal function, but the very same symptoms in an elderly person may be ascribed to 'natural ageing'. Post-mortem examination of the brain of both individuals reveals senile plaques and neurofibrillary tangles, although they may differ in quantity and distribution. Also, similar changes are seen in Down's syndrome at a much earlier age (see 'Premature ageing syndromes', below). In reality, the impairment of normal brain function is comparable in all these cases. It has been said that Alzheimer's disease is not closely associated with ageing, because so many very old individuals retain normal intelligence and brain function. Nevertheless, it is uncommon for such individuals to retain the normal learning ability and the memory that they had when much younger. The reasonable conclusion is that brain function declines during ageing, but at a faster rate in some individuals than others. Alzheimer's disease only becomes a specific condition when it occurs in late-middle-aged individuals, who are otherwise normal. The onset of the disease can be familial, and there is much speculation that environmental agents may be responsible for other sporadic cases.

The major histological features of Alzheimer's disease are neurofibrillary tangles, neuritic or senile plaques, and the accumulation of amyloid, both within the plaques and also in intracortical arteries. The neurofibrillary tangles are bundles of paired helical filaments in the cytoplasm of neurones that encircle or displace the nucleus. β-amyloid precursor protein (APP) has been the focus of many molecular studies (reviewed by Tanzi, St. George-Hyslop & Gusella 1991; Bush, Beyreuther & Masters 1992; Hardy & Mullan 1992). APP is a glycosylated transmembrane protein with a small cytoplasmic domain, a transmembrane domain and a large extracellular domain. The chief component of amyloid in senile plaques and cerebrovascular deposits is the βA4 peptide fragment, about 40 amino acids in length, which is part of the APP sequence. About one-third of the βA4 sequence is in the transmembrane part of the APP protein, and the remainder in the extracellular region. Abnormal processing of the APP protein produced the βA4 fragment, and if large amounts of this are produced, then amyloid plaques occur. The problem is then to explain the abnormal processing. One clue is the fact that the APP gene resides on chromosome 21, and Down's syndrome patients (trisomic for this chromosome) invariably develop Alzheimer's disease. Therefore excess production of APP may trigger abnormal processing. It is known that the APP protein is also processed to produce normal amyloid β-peptides in cultured cells and body fluids. The relationships between normal and abnormal processing are far from clear. Studies of the familial form of Alzheimer's disease show that a gene segregates on chromosome 21, but this is separable from the APP gene by recombination. Other studies on the familial disease have detected mutations in the APP sequence, most of which are associated with β-amyloid deposition (Hardy 1992). These mutations may bring about abnormal processing. However, many families with early-onset Alzheimer's disease have the normal sequence in the entire open reading frame. Cellular ageing in many contexts is associated with the accumulation of insoluble proteinaceous material, and there has been much discussion of the possibility that proteolysis is defective in aged cells and tissues. The accumulation of β-amyloid in plaques may be due not only to abnormal processing, but also to a failure to degrade the βA4 peptide as it is formed. Whatever the molecular origins of the tangles and plaques in Alzheimer's disease, the end result is the loss of normal neurones, which cannot be replaced.

Other brain pathologies are also age-related. Parkinson's disease occurs from about age 50 onwards and is progressive, with the primary effect on motor function. Fortunately, a much smaller proportion of individuals are affected than is the case with Alzheimer's disease. More generalised age-related changes are seen in the neuroendocrine system. For example, loss of cells in the hypothalamus affects hormone production and therefore the normal function of target organs. Elderly individuals often develop hypertension and they may have impaired glucose tolerance, both of which are likely to be directly related to abnormalities in hormone levels.

The brain is the most complex organ in the body, and it will be a long time before its structure and functions are understood. It is therefore perhaps unlikely that current research on a specific brain disease, such as Alzheimer's, will soon lead to effective treatments. Nevertheless, the research will certainly increase our understanding of biochemical abnormalities in neurones and other brain cells. This in turn will expand our knowledge of the molecular changes that can accompany ageing. In combination with studies on other cell, tissue and organ systems, much will be learned about processes that are potentially reversible and those that are irreversible.

The retina can be regarded as an extension of the brain. The two types of photoreceptor cells (cones and rods) transmit signals to the bipolar cells and thence to ganglion cells with axons that extend to the tectum of the brain. Photoreceptor cells are adjacent to a layer of pigmented epithelium attached to Bruch's membrane. Oxygen is supplied by an extensive capillary network. Several features of the retina illustrate the inevitability of age-related changes. The photoreceptor elements in the rods are continually turned over. New receptors are added at the top nearest the light source, like adding a new coin to an existing pile, and old receptors at the base are removed. About 90 new receptor elements are added per day in each rod in the retina, and of course a similar number are removed. Thus, each receptor element has a finite lifetime. Removal of the elements at the end of their useful life depends on lysosomal activity. Most protein is degraded, but some is insoluble and gradually accumulates in secondary lysosomes. The underlying epithelium has an active role in taking up the spent material by phagocytosis and degrading it. However, a steady state is not achieved, and with time increasing amounts of debris accumulate. This leads to a thickening of Bruch's membrane, and as a result there can be areas of detachment of the retina from the membrane. The macula is the central and most sensitive part of the retina. Degeneration of the macula is age-related, and this is largely due to defects in the epithelial membrane. In the region of atrophy of the epithelium, the photoreceptors degenerate, leading to blindness. The retina is also subject to other damage, resulting from age-related changes. Diabetes, including the late-onset Type II disease, results in abnormal sugar levels in blood, and this eventually brings about retinopathy. About 60% of diabetic patients develop retinopathy 15–20 years after the original diagnosis. Hypertension and arteriosclerotic changes in the retinal vessels, both of which are age-related, also severely affect normal retinal function.

The lens of the eye also illustrates some basic features of ageing. The various crystallin proteins of the lens are produced by cells that subsequently degenerate to form the transparent and somewhat elastic lens structure, surrounded by lens epithelium. The loss of accommodation of the lens is one of the best biomarkers of ageing. It is due to the gradually reduced elasticity of the crystallins and the failing ability to focus on both near and distant objects. The crystallin molecules are never replaced, and they are therefore inevitably subject to intrinsic, irreversible changes. These may include non-enzymic glycosylation, oxidation, partial denaturation, cross-linking of molecules, racemisation of amino acids, and so on (see Chapter 4). As these changes continue, a loss of transparency occurs with the clinical manifestations of cataracts. There are several causes of cataracts, but senile cataracts are the commonest category. The different categories of cataract depend mainly on the position and the colour of the opacity, and there may also be abnormalities in the lens epithelium. Cataracts provide one of the best examples of an age-related change that can be successfully treated by surgical intervention.

The ear is also subject to irreversible age-related changes, which are mainly brought about by loss of cells (presbycusis). Loss of sensory hair cells responding to the highest sound frequencies is the earliest change. It is well known that children can hear higher frequencies than adults, and this change is progressive with age. In addition, cells are lost in the auditory pathway from the receptor cells of the cochlea to the auditory cortex of the brain. There may be structural changes in the middle ear, such as stiffening of the tympanic membrane, or progressive changes in the ossicles or other bony structures. There can also be atrophy of the spiral ligament of the ear. Not all changes simply lead to loss of hearing, as ageing can produce other malfunctions. For example, the axons to the brain can be continuously stimulated in the absence of sound, which results in a high-pitched ringing in the ears (tinnitus). The ear provides a notable example of a highly specialised structure, with extremely limited capacity for repair of damaged components.

THE VASCULAR SYSTEM

With regard to the effects of ageing, the arteries can be broadly assigned to two categories. The smaller arteries, which are potentially replaceable, and the larger arteries, which last the whole lifetime. When tissue is damaged, bleeding ceases when the blood forms a clot, and repair of the damage involves vascularisation, that is, the formation of new small arteries and veins. The effects of ageing are seen primarily in the large arteries, which cannot be replaced in the same way. The structure of the large artery walls is complex and can be divided into three parts: an inner layer of endothelial cells with underlying connective tissue, consisting of collagen, proteoglycans, elastin and a number of other matrix glycoproteins; a middle layer of smooth muscle cells, together with elastic fibres, and external to this another elastic layer of connective tissue, in which there are embedded nerve fibres and small blood vessels to provide oxygen and nutrients. An essential feature of arteries is their elasticity and resilience, which are necessary to accommodate the changes in pressure produced by the pumping of the heart. During ageing, elasticity declines. The connective tissue layers become less flexible because long-lived molecules such as collagen and elastin eventually become cross-linked. As a result of the loss of elasticity, the vessels expand less readily and blood pressure may increase. Arteriosclerosis comprises a group of related abnormalities, all characterised by thickening and hardening of arterial walls. It is probable that the glycation of proteins is one important cause of the thickening of artery walls, since the condition is common in diabetics. Any reduced blood supply has secondary effects, particularly on the function of kidneys, which normally receive about 25% of cardiac output. It is common for renal function to decline by as much as 40% during ageing.

In the extensive research on heart disease, most attention is paid to atherosclerosis, which is caused by the accumulation of plaques on the inner layer of the larger arteries. These plaques have a core of extracellular lipid and cholesterol, with lipid-laden 'foam cells' derived primarily from macrophages. The core of the plaque is covered with a fibrous cap consisting of proliferating smooth muscle cells, collagen, elastin and proteoglycans, together with other cell types, such as lymphocytes and macrophages. The origin of plaques is, of course, the subject of intense debate, but it is widely believed that they arise as a result of injury to the arterial endothelium. The integrity of this endothelium is essential for the normal structure and function of the vessel wall. The endothelium acts as a semi-permeable membrane, which controls the transfer of molecules into the arterial wall and has several other physiological functions, including modulating blood flow and serving as a barrier between blood and the thrombogenic subendothelial layers. There is evidence that the atherosclerotic plaques are clonal in origin (Benditt & Benditt 1973), and they probably arise from smooth muscle cells migrating from the deeper layers of the arterial wall. They proliferate in response to platelet-derived growth factors following adherence of platelets to areas of damaged endothelium. The lipid in the plaques may become oxidised, and this sets up an inflammatory reaction in the plaque leading to complications such as increase in size and ulceration.

Stiffening of the arteries (arteriosclerosis) is an almost universal age-related change, but atherosclerosis is strongly influenced by diet and life style, although its prevalence increases with age. It is clear from the extensive studies of vascular disease that the large arteries are unable to sustain indefinitely their normal structure and function. The continual mechanical stress to which they are subjected is associated with abnormalities in the connective tissue components, particularly loss of elasticity, and also in the endothelial and sub-endothelial layers where the plaques form. This gradual loss of normal structure has several consequences that can be lethal. Damage to the endothelium results in a release of thrombogenic factors, the formation of blood clots and the possibility of essential arteries becoming blocked. The increasing number and size of plaques will also eventually result in the blocking of normal blood flow. Loss of elasticity and high blood pressure can result in dilation of the blood vessel (aneurysm) and the possibility of rupture, haemorrhage and stroke.

It has been said that no mechanical pump has the efficiency, durability and reliability of the human heart, but no pump can continue indefinitely, unless worn-out parts are replaced. The anatomical design of the mammalian heart does not have an inbuilt capacity to replace such parts, and it necessarily has a limited lifespan. A characteristic feature of the aged heart is loss of compliance, or stiffening of the wall, due at least in part to cross-linking of muscle fibres. The causes and consequences of heart failure are highly complex and obviously cannot be reviewed here. Heart function may fail if the supply of oxygen to the muscle cells is severely diminished, for example, by blockage of the coronary artery. This failure may be gradual over a long period of time, or it may be sudden. Loss of oxygen reduces cardiac output, and if severe, causes irreversible damage, or necrosis, of heart tissue. Unfortunately this tissue has very limited powers of regeneration and repair. Heart muscle cells are post-mitotic and, unlike muscle in other parts of the body, there is no pool of myoblasts that can divide and differentiate to replace lost cells. Components of the heart, such as valves, are also subject to gradual failure after prolonged function. The valves are a target for chronic calcification, which is common in the aged.

In the Introduction the analogy of machines with limited life was discussed. A machine can be kept functional if parts are replaced, but to do this the machine is first stopped. In the case of the heart, continual function is essential for survival of the organism, and this severely limits the possibility of ongoing repair. There is also a potential instability in heart function, which is in part a feedback mechanism: if oxygen supply for any reason declines, then cardiac output is impaired, so oxygen supply declines even further. This can only lead to irreversible arrest.

RENAL FUNCTION

The kidney is far less sensitive than the brain or heart to tissue damage, since its function is maintained even if about two-thirds of its mass is damaged by injury. Nevertheless, continuous physiological and anatomical changes occur with age, and failure of function is common. There is a strong interdependence between the vascular and renal system. The kidney receives about 25%

of cardiac output, and about 1700 litres of blood are processed per day. The glomeruli of the kidney are sensitive to vascular abnormalities such as hypertension, and the kidney itself has an active role in the control of blood pressure. This is in part due to its role in controlling the level of sodium ions in the blood, and also to the secretion of hormones such as renin and prostaglandins. Thus, the kidney plays a vital role in the homeostatic mechanisms that control water balance and the quality of the blood supply.

With ageing, the number of glomeruli declines, and those that remain may become abnormal with thickening of membranes. Renal arterioles may be replaced by collagen, which is itself subject to age-related cross-linking. Sclerotic and fibrotic changes occur in blood vessels, and these are in large part responsible for the senile changes seen in the kidney. All these anatomical and histological signs of deterioration are associated with a generalised loss of physiological function, including reduced ability to conserve sodium when it is depleted in the diet, and adverse effects on the control of blood pressure. It is significant that glomeruli filtration rate is maintained until the age of 40, but thereafter declines at a rate of about 1% per year, albeit with considerable variation between individuals (Wesson 1969). The young adult kidney has considerable reserve capacity, but nevertheless with time these reserves are used up. Normal function may be maintained until death from other causes, but commonly the failure of the kidney to replace lost glomeruli, or repair other intrinsic damage, leads to renal failure with diverse physiological side effects.

LATE-ONSET DIABETES

Diabetes affects carbohydrate, fat and protein metabolism, primarily as a result of abnormally high glucose in blood and tissues (hyperglycaemia). There are two major types of diabetes. Type I is of early onset, and is due to the loss of the Islets of Langerhans in the pancreas, which secrete insulin. We are concerned here only with Type II diabetes, which is strongly age-related and also much more common than Type I. The causes of Type II diabetes are much less well understood, and as well as intrinsic factors, including genetic components, there are strong environmental influences, such as a diet leading to obesity. The primary physiological defect is not established. It could be a failure to produce sufficient insulin in response to the sugar load, the existence of defective receptors on target cells, abnormalities in the processing of the signal from the receptor, or any combination of the three. The end result is essentially the same, that is, hyperglycaemia, which results in several important pathological changes. Two major mechanisms linking hyperglycaemia to subsequent tissue damage have been proposed. The first is the non-enzymic glycosylation of proteins, which results in the irreversible production of advanced glycation end-products (see Chapter 4) at various locations in the body. The likely significance of non-enzymic glycosylation is shown by the fact that the severity of the course of the disease is monitored by the measurement of levels of glycosylation in red cells. The second is based on the possibility that an increase in intracellular glucose would be followed by its conversion to sorbitol and fructose. The increased sorbitol and fructose could then lead to osmotic injury and also to a decrease in inositol, which in its various phosphorylated forms has a fundamental role in signal transduction in cells.

One of the most consistent morphological features of diabetes is the diffuse thickening of basement membranes. This is seen in capillaries of the skin, muscle, retina and kidney. Complications include generalised atherosclerosis, renal failure, cataracts, defects in the retina (retinopathy), in peripheral nerves or other components of the central nervous system. It is very probable that advanced glycation end-products are an important component of these pathological changes. It is, for example, hardly likely that defects in intracellular sorbitol metabolism would bring about changes in the crystallin proteins of the lens.

Diabetes well illustrates the fundamental consequences of what might be initially a simple molecular defect. Suppose an age-related abnormality in protein metabolism reduces the normal number of insulin receptors. The resulting hyperglycaemia can then have disastrous consequences on the vascular system, vision, kidney function and so on. Thus at one level there might be a primary age-related defect, but there are subsequently multiple pathological side effects that are also age-related. It is well known that there is in the human population a continuous spectrum from severe Type II diabetes, through to moderate, mild or none. All this is consistent with a stochastic occurrence of defects that can undermine the normal metabolic control of a hormone, which itself plays a central role in homeostatic mechanisms.

OSTEOARTHRITIS AND OSTEOPOROSIS

The ends of bones in joints are capped by articular cartilage, which serves as an elastic shock-absorber and wear-resistant surface. The surfaces are lubricated by fluid produced by the synovial membrane, which lines the parts of the joint that are not load bearing. The load-bearing regions of joints do not have nerves or blood vessels, but the cartilage is a metabolically active tissue. Chondrocytes synthesise the collagenous matrix, consisting mainly of collagen Type II and proteoglycans, and they also produce collagenase and other proteases, which allows turnover of collagen. Protease inhibitors also have a role in cartilage metabolism. Thus the cartilage of joints is not just a mechanical structure, but has considerable powers of turnover, replacement and maintenance. Unfortunately, these are unable to cope with prolonged use, and as a result 80–85% of the population over 70 are affected by osteoarthritis.

This disease is due to endogenous changes in the joints, particularly those that are weight bearing. The earliest changes are loss of proteoglycans and also chondrocytes, which are essential for cartilage maintenance. There is probably an imbalance between the synthesis and removal of cartilage components. As a result, the articular cartilage becomes pitted or fissured, and pieces may flake off. This is associated with inflammation of the synovial membrane. As in many age-related changes, once an apparent steady state is lost, the problem of maintenance escalates. In this case, the continued death of chondrocytes and damage to the synovial membrane leads to the release of degradative enzymes, cytokines and possibly free radicals. As the cartilage becomes more eroded, bone may be exposed, and this becomes vascularised, thickened and also polished by continual joint motion. All these changes result in malformation of the normal contours of the ends of the bones. The ageing of joints has something in common with that of teeth: both structures are complex, with mechanical as well as cellular components, and both are subject to wear and tear that is essentially irreversible.

Osteoporosis, another important age-related disease, is due to a reduction in bone mass. This commonly results in fractures, as bones become fragile and brittle. Bone structure is normally maintained by a balance between formation and reabsorption, but after the age of about 30 reabsorption gradually begins to outpace formation, so there is a gradual decline in bone mass. This decline accelerates in females after the menopause, and as a result osteoporosis is commoner in women (post-menopausal osteoporosis) than in men. Nevertheless, it commonly occurs in the later decades in men and is then referred to as 'senile osteoporosis'. As in the case of many age-related diseases, no obvious line can be drawn between normal bone loss and the increased loss in osteoporosis, which then greatly increases the likelihood of fractures. The causes of osteoporosis are unknown, but it is agreed that there is an imbalance between the formation and loss of bone. It is also clear that hormonal changes are involved, which in turn relate to calcium metabolism. Reduced physical activity also results in bone loss, and genetic factors are known to be important.

SKIN

One of the most obvious of age-related changes is the appearance of skin. Our judgement of a person's age is in large part determined by the gradual alteration of skin, particularly on the face and hands, or other exposed parts of the body. Young skin is smooth and resilient, but as it ages this resilience is lost with an increase in sagging and wrinkling. The gradual transformation is due to both cellular and extracellular changes. One of the major alterations is due to the loss of organisation of extracellular collagen. Collagen becomes increasingly cross-linked with age; fibres thicken and they are in somewhat disorganised ropelike bundles, rather than in the orderly pattern seen in young skin. There is decreased epidermal renewal and tissue repair, and a slower rate of wound healing. The number of cells in the dermis, including fibroblasts, macrophages and mast cells, decreases during ageing, and there is decreased vasculature. This makes aged individuals much more sensitive to hypothermia than young ones. Older people are also more susceptible to hyperthermia, due to a decline in the number of sweat glands declines with age, and this may increase mortality rates during heat waves. Hair loses pigmentation owing to loss of melanocytes. This clearly shows the random stochastic nature of the events that characterise ageing, since cells in individual follicles are affected independently.

Both keratinocytes and fibroblasts have finite cell proliferation (see Chapter 5), and it has been demonstrated that there is an inverse relationship between the age of donor and the number of population doublings of their skin fibroblasts. The division potential of cells from individuals 80–90 years old is about 60% of those from children (Martin et al. 1970). This means they still have very significant remaining growth potential, but there may be more non-viable cells or cells with very limited growth, which could impair wound healing. For example, ulcers may occur on the ankles of old people, which do not readily heal. It is probably significant that patients with the inherited premature ageing disease Werner's syndrome (see 'Premature ageing syndromes', below), are much more prone to non-healing ulcers, and their skin fibroblasts have very limited *in vitro* growth potential.

The ageing of skin in exposed parts of the body in non-pigmented individuals is in part due to the UV light component of sunlight (especially UVB). A common feature is the appearance of 'age spots', which are areas of increased pigmentation probably due to clonal growth of affected keratinocytes. It is possible that mutations or epimutations that give rise to clones of altered cells are an important component of ageing, not only of skin, but of other tissues as well. Basal cell carcinomas induced by exposure to sunlight become much more common after middle age. Recently it has been shown that the ability to repair UV-induced DNA damage in skin significantly decreases with age (Wei et al. 1993).

The structure of the epidermis is continuously renewed from a stem line of epidermal cells. Indeed, at first sight it seems that the skin might have the capacity to renew or replace all component parts, with the potential for indefinite survival. It is all too obvious that the biological reality is otherwise, and that the intrinsic structure and functions gradually decline with age.

THE IMMUNE SYSTEM

Most of the important components of the immune system are derived from the stem cell population of bone marrow. Thus, unlike other tissues and organs discussed in this chapter, the immune system might be expected to be capable of continual renewal. It is significant that this is not in fact the case, because it indicates that even dividing cells *in vivo* are subject to ageing, just as are dividing cells *in vitro* (see Chapter 5). This has been demonstrated in the case of spleen cells, which have a specific immunological memory. Transplantation experiments show that these cells are capable of about 100 divisions *in vivo*.

The decline of the efficiency of the immune system during ageing is well known. There is a progressive qualitative and quantitative decrease in the capacity to produce antibodies. There is also a steady decline in T lymphocyte function. Abnormal aggregates of lymphocytes may appear in bone marrow and elsewhere during ageing. Although generalised auto-immune reactions, such as those seen in lupus erythematosus, are not age-related, some other auto-immune diseases occur much more commonly in old people. Giant cell arteritis is restricted to people over 65. Auto-immune reactions during ageing are not unexpected, since we know that abnormal proteins are produced during ageing and some of these will be distinct from self antigens and therefore provoke an immune response. The decline of immune function has been examined in most detail in experimental animals. The results show that the reduced efficiency of B and T lymphocyte populations is due to intrinsic changes in the cells themselves, rather than to extrinsic influences (Makinodan 1979; Makinodan & Kay 1980). The molecular basis of these intrinsic changes is not known.

The thymus is a defined organ of the immune system, and it plays a central role in cell-mediated immunity. Immature T lymphocytes from the bone marrow migrate to the thymus where they differentiate into mature T cells, before entering the vascular and lymphatic systems. The thymus also secretes a number of protein factors that strongly influence T cell function. It is sometimes regarded as a useful model system for the study of ageing, because it appears to follow a defined programme of age-related changes. In humans it reaches its maximum size at puberty and thereafter undergoes a progressive decrease, and by age 50 is less than 15% of its maximum weight. Presumably its relatively short active lifespan is sufficient to provide T cell–mediated immunity for the whole lifespan. The thymus can therefore be regarded as yet another example of an organ with an evolved design that is sufficient for the 'disposable soma' of the species.

CANCER

Cancer is the result of breakdown of the normal controls of development and tissue integrity, and the resulting cell growth can be benign or malignant. Cancers are of clonal origin, and the phenotype of the cells is heritable. In the emergence of malignant tumours both the initial and subsequent events are believed to depend on changes at the DNA level, which determine the ultimate cell phenotypes. Although mutations in oncogenes and tumour suppressor genes are well documented (Levine 1993), there may also be other events, such as epigenetic changes, that are heritable and have important effects on cell phenotypes (see Note 8.1). Some cancers occur commonly in childhood and develop rapidly, such as leukaemias and brain tumours, but most carcinomas are age-related. It is not appropriate to summarise here the features of the many different carcinomas or other neoplasms. Although these cancers are age-related, their distribution in different populations indicates that many are caused by environmental factors, the most clear-cut example being smoking and lung cancer. Epidemiological studies of this disease indicate that the average time for the initiation of the event to the diagnosis of the disease is about 20 years. Moreover, the increase in incidence with age gives a slope on a log–log scale which indicates that four to six successive events are necessary before the malignant tumour is detected. The slope is similar for most carcinomas in adults. Curiously, the fact that several sequential events are necessary has been interpreted to mean that these cancers bear no relationship to ageing per se. The title of one paper on the subject is 'There is no such thing as ageing, and cancer is not related to it' (Peto, Parish & Gray 1986). The interpretation is that the slow 'multiple-hit' progress of the disease inevitably means that it is more common in elderly people. This seems to reveal a misunderstanding of the nature of ageing, since it is also, at least in some sense, a temporal, multiple-hit process. People die when they are old because many events have occurred in their cells, tissues or organs. The confusion arises because the probability of each event in tumour progression is not thought to be related to the age of the individual in which it occurs. In other words, the events that give rise to tumours are not more likely to occur in old people than in young. At least that is the supposition, but the alternative view is that the age-related changes in the human organism actually increase the probability of tumour appearance and growth (see Anasimov 1986, 1991). Initial results on carcinogen-induced skin cancers in mice indicated that the sequential events that occurred were independent of the age of the animal (Peto et al. 1975). A later, more thorough study that was, at least in part, designed to demonstrate the fundamental difference between tumour progression and ageing, gave an equivocal result. The prediction was that the initiation of carcinogenesis and the subsequent progress of the disease would be independent of the age of the animal, but this was not in fact the case (Stenback, Peto & Shubik 1981). An interesting recent study shows that the DNA repair capacity of human skin declines with age (Wei et al. 1993). The authors suggest that the increased risk of UV light–induced skin cancer, which begins in middle age, may be related to this decline.

In the context of ageing, the comparison between humans and mice or rats is revealing. The likelihood of a 2–3-year-old rat developing a tumour is broadly equivalent to the likelihood of a 60–70-year-old human developing the same or a similar kind of tumour. This is shown in Figure 8.1. In both cases the incidence rises at about the fifth power of the age. If the same number of events are occurring during tumour progression in the two species, it

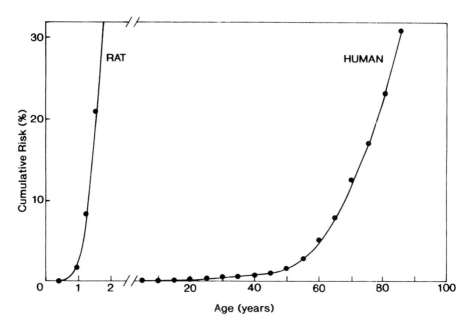

Figure 8.1. The age of onset of carcinomas in the rat and humans. The rising incidence is approximately the fifth power of the age. (Reproduced with permission from Ames et al. 1985.)

can be calculated that the individual events in mouse cells are occurring 10^9 times as frequently as in human cells (Peto 1977). This vast difference would be less if, for instance, there were six events before the emergence of human tumours, and only two or three in mice. It is not known whether or not this is the case, but it is known that normal rodent cells in culture undergo neoplastic transformation at a very much greater frequency than normal human cells, either spontaneously or after carcinogen treatment (see Chapter 5). Overall, the observations on cells and animals show that the human cells and tissues are much better protected against the development of cancer than are mouse tissues. Yet when the frequencies of mutations and/or chromosome abnormalities in the two species are measured, there are not obvious differences. Thus, the common view that tumour progress depends on a combination of gene mutation and cytogenetic abnormalities does not explain the fundamental differences between short-lived and long-lived species. Instead one should perhaps look at the stability of epigenetic controls. A real possibility exists that human cells are better 'locked in' to their appropriate phenotype, and with regard to tumour formation survive damage to DNA very much more effectively than do rodent cells. Evidence already exists that reactivation of the inactive X chromosome in mice occurs spontaneously during ageing, whereas in humans it seems to be a very rare event (see Chapter 4). It can be concluded that the emergence of cancers is due to the eventual breakdown of the normal controls of cell phenotype. These controls constitute a maintenance mechanism, which is much more effective in long-lived species than short-lived ones. The eventual failure of this maintenance, which leads to cancer, is one important component of ageing.

PREMATURE AGEING SYNDROMES

The vast resources that are assigned to human health care have resulted in the identification of over 3000 different heritable abnormalities. A small subset of these have specific effects on ageing. Current information was reviewed by Martin (1978), who concluded that the number of known inherited conditions that in one way or another affected ageing, or age-related pathologies, was steadily increasing with time. This suggests that many different genes affect ageing. However, many of these inherited conditions are related to the premature occurrence of a particular age-related pathology, and we are concerned here with genetic defects that have a wider spectrum of premature ageing effects. Martin assessed these in relation to the degree of similarity of the premature ageing seen to be the natural ageing of normal individuals. None mimicked such ageing completely, and he therefore coined the phrase 'segmental progeroid syndrome' to describe those genetic conditions that accelerate several or many features of ageing, but not all of them. He ranked these according to their similarity to natural ageing. Down's syndrome was ranked first, followed by Werner's syndrome and progeria (Hutchinson–Gilford syndrome). (Others near the top of the list were ataxia telangiectasia and Cockayne's syndrome, which are better known for their defects on DNA repair. Also, children with Bloom's syndrome nearly always develop tumours and have a short expectation of life, so it is difficult to know if they would exhibit premature ageing in, say, their fourth or fifth decades.)

Down's syndrome is caused by a trisomy of chromosome 21 (or a translocation involving chromosome 21), so the phenotype is a result of a defect in gene dosage. Major features are significant mental retardation and the invariable early development of Alzheimer's disease. The pathological changes seen in the brain are identical to those seen in patients with Alzheimer's disease who have a normal karyotype. Many individuals with Down's syndrome are born with heart defects, and most die from heart disease. Down's syndrome children also have about a 20-fold increased risk of acute leukaemia and have a reduced resistance to infections, especially of the respiratory tract. There is an increased deposition of lipofuscin in tissues, some premature greying or loss of hair and increased susceptibility to diabetes. Studies of translocations have defined the critical region of chromosome 21, which is responsible for the phenotype, and molecular procedures are being used to identify a specific gene in this part of the chromosome. One of significant interest codes for superoxide dismutase (SOD), an essential enzyme in the defence against free radical damage. A 50% increase in this enzyme might disturb the normal balance with catalase and glutathione peroxidase and therefore lead to an increase in hydrogen peroxide and hydroxyl radicals (see Fig. 3.1).

The gene causing progeria is inherited as an autosomal dominant that fortunately is extremely rare. Infants are outwardly normal, but after a few years children develop a progressively aged appearance, and generalised atherosclerosis is the cause of death, usually well before the age of 20. Other features are a very short stature with spindly arms and legs, baldness and loss of subcutaneous tissue. However, the children are intelligent, and there is no evidence of brain abnormality.

Werner's syndrome is caused by an autosomal recessive mutation and is perhaps the most interesting of the premature ageing diseases (reviewed in Epstein et al. 1966; Salk, Fujiwara & Martin 1985). It is far less extreme than progeria, and affected individuals are normal for 20–30 years, apart from premature whitening of the hair. Thereafter progressive changes occur, including cataracts of the juvenile type, diabetes, loss of subcutaneous fat on arms and legs, osteoporosis, and an increased incidence of neoplasia. Affected individuals in their forties have a prematurely aged facial appearance. Studies of cells demonstrate chromosomal instability (Salk et al. 1981), and more recently it has been shown that the mutation frequency is also elevated (Fukuchi et al. 1985, 1990; reviewed by Monnat 1992). Cultured fibroblasts also have increased levels of a heat-labile fraction of two enzymes tested (Holliday et al. 1974; Tollefsbol et al. 1982). In addition, these cells grow slowly and have very limited growth potential (Martin et al. 1970; Thompson & Holliday 1983). This may be clinically significant, because Werner's patients often develop ulcers on their ankles, which frequently do not heal. The brain is unaffected, but atherosclerosis and other vascular defects develop prematurely and are the usual cause of death at an average age of 46 years. Although Martin considered Down's syndrome to be the best candidate for a 'segmental progeroid syndrome', these individuals are abnormal from birth, and frequently have a heart defect. Individuals with Werner's syndrome, on the other hand, are completely normal until about the age of 20.

In spite of considerable investigation, the biochemical basis of the defect is unknown. So far there are only a few clues. It has been reported that the activity of the enzyme 5-hydroxymethyl uracil (HMU) glycosylase is significantly reduced (Ganguly & Duker 1992). This is one of many enzymes that remove damaged bases from DNA (see Chapter 3). HMU is produced from thymine by oxygen free radical attack, so a reduction in the enzyme might explain the increased mutation rate and susceptibility to cancer in Werner's syndrome patients, but is perhaps less likely to be responsible for chromosomal abnormalities. The other observation shows that hyaluronic acid metabolism is abnormal in Werner's syndrome patients, since they excrete an elevated amount in urine. Whatever the defect, it is remarkable that a single recessive mutation, presumably causing a loss of function, can have so many diverse

Table 8.1. *Relationships between cell or tissue maintenance and human age-related disease*

Failure of maintenance	Major pathologies
Neurones	Dementias
Retina, lens	Blindness
Insulin metabolism	Type II diabetes
Blood vessels	Cardiovascular and cerebrovascular disease
Bone structure	Osteoporosis
Immune system	Auto-immune disorders
Epigenetic controls	Cancer
Joints	Osteoarthritis
Glomeruli	Renal failure

effects on the phenotype. It is hard to see how a single biochemical defect can affect, for example, proteins of the lens of the eye, DNA metabolism and atherosclerosis. This is discussed further in Chapter 9.

CONCLUSIONS

This chapter has reviewed specific aspects of human ageing and age-related disease. I asserted in Chapter 1, and re-emphasise here, that the distinction between so-called natural ageing and the pathologies that are common in old people is artificial. What we see is an increasing likelihood of many diseases in individuals as they age, which does not, of course, mean that all individuals develop all the pathologies. The point was made in Figure 1.4. For the gerontologist the problem is to explain how diverse changes in many tissue and organ systems occur with some degree of synchrony. The teeth, lenses, blood vessels, joints, hair and so on can all last a lifetime, but all decline in their structure and/or function. Maintenance is essential for ongoing function, usually for a considerable proportion of the lifespan, but for the reasons outlined here and in Chapter 2, maintenance cannot prevent ultimate decline. The relationships between failure of maintenance and particular diseases are listed in Table 8.1. Maintenance must be most effective in long-lived animals, but is never effective enough to prevent eventual deterioration. Two very important issues are the modulation of the efficiency of maintenance in relation to lifespan, and the number of genes involved in this modulation. These are discussed in the next chapter.

9

A better understanding of ageing

It is indeed remarkable that after the seemingly miraculous feat of morphogenesis a complex metazoan should be unable to perform the much simpler task of merely maintaining what is already formed.

– George C. Williams

AGEING AND DARWINISM

The above quotation by Williams pinpoints the fundamental issue, which is discussed in previous chapters, and the answer takes us a long way towards a basic understanding of ageing. As in so much of biology, the explanation for ageing can only be found in its evolutionary origins. The universal feature of living organisms is their ability to reproduce. For reproduction to be effective many more offspring are produced than in an ideal environment would be necessary to replace parents. The natural environment is never ideal: there is competition between species for resources, especially food. There is the ever-present competition between complex organisms and less complex parasites and pathogens. There is competition between plants and animals, especially the synthesis of toxic products in plants by secondary metabolism. A similar chemical warfare is common between small invertebrates, such as insects, and their predators. As well as competition between species, there is competition between members of the same species. All this was recognised by Darwin, and forms the basis of his theory of natural selection. In a competitive world, the best-adapted individuals are more likely to survive than the less well adapted. Fitness, in the terminology of the population geneticist, is the likelihood of producing offspring that will themselves produce offspring. In a sexually-reproducing population in which the number of individuals remains roughly constant, each female should produce on average two reproducing adults; but fitness varies, so the fittest parents will have a greater probability of contributing to the next generation than the less fit. To achieve this end, animals and plants overproduce offspring. This usually means that only a minority of offspring survive to adulthood and contribute to the following generation. Most die from predation, starvation, disease or some other hazard of the environment. This also means that the expectation of life of a population of offspring is short, and also, as Medawar (1952) first clearly stated, that few will survive for a long time.

This provides the answer to the problem Williams raised. Evolution has indeed produced the apparently miraculous feat of morphogenesis and the reproducing adult, but what is the advantage of preserving the adult indefinitely, when such preservation will be cut short by the natural hazards of the environment? Clearly, animals must survive long enough to produce adequate numbers of offspring, because that is the measure of fitness; but to evolve the potential for indefinite survival, that is, survival without ageing, is a totally different issue. On the basis of very simple and reasonable assumptions, it can be shown that the ability of complex organisms to survive indefinitely would necessarily decrease fitness (see Fig. 7.1). This is not the case for simple organisms. Many microbes propagate themselves by binary fission, simple animals such as *Hydra* bud off new offspring and many plants have the ability to reproduce vegetatively, that is, without sexual reproduction. These life forms have a plasticity that makes them potentially immortal. They still frequently die from the hazards of a natural environment, but they do not necessarily die from ageing.

In the evolution of complex animals, the feature that dominates all others is the successful production of offspring. Fitness is increased if resources are allocated to reproduction and reduced if they are allocated to the long survival of parents. This provides the explanation for ageing. There is a trade-off between reproduction and long-term survival, and evolution necessarily dispensed with the latter. In other words, the germ line is potentially immortal, but the soma is disposable. In the evolution of animals, this began very early on. Adult nematode roundworms have a fixed number of post-mitotic cells, and they have a clearly-defined lifespan. Almost all animals more complex than coelenterates and flatworms have a structure that is incompatible with continuous survival. Amongst vertebrates, some that continue to grow as adults (for example, some large fish and reptiles) survive for a long time, but even their extended lifespan is a minute fraction of evolutionary time.

IMMORTALITY

In Chapters 2 and 8, some features of the mammalian organism were reviewed that inevitably contribute to ageing. It is instructive to consider what a complex organism such as the mammal would need to allow it to achieve an equilibrium or steady state, with potential for indefinite survival. It is fairly easy to imagine renewable tissues such as skin and blood, where cells turn over and are continually replaced. There would also have to be a turnover of long-lived proteins, such as collagen, which otherwise lose elasticity from cumulative cross-linking. Some organs, such as the liver, already have powers of regeneration, and it is not difficult to envisage age avoidance mechanisms in several others. This could also apply to muscles and most connective tissue, although joints might provide special problems for ongoing repair mechanisms. Teeth could be replaced by growth at the base or by growing new teeth, just as adult teeth replace early milk teeth. However, the evolved design of the heart and vascular system make it very hard to see how cell replacement and repair could be effective. Instead, it would probably be necessary to have one functional system in place, whilst an active morphogenetic process is producing a replacement. The new system would take over before the old has ceased function, and the latter could then be autolysed and the components reutilised. The scenario is even more fanciful when the preservation of an efficient central nervous system is considered. Neurones are postmitotic, long-lived cells. They are also the seat of learning and memory. Birds that sing a new song each season are reported to replace some of their brain cells, and one could imagine neurone replacement for many basic functions. When mammals evolve efficient memories, however, it becomes very difficult to see how memory could survive cell replacement. Of course, we do not understand the cellular and molecular basis for memory, so any proposal for the design of an everlasting memory is necessarily fictional. Nevertheless, an advanced and complex brain capable of indefinite survival might have evolved had there been an essential need for it. As it is, we have a brain that falls very far short of this.

REPLACEMENT OF PARTS

In modern medical practice much attention is paid to various techniques of organ transplantation. Kidney transplantation is common, and heart transplants are no longer unusual. Liver and lungs are also sometimes replaced. The surgical procedures have been developed, but the prevention of rejection of the foreign tissue is still a major problem. All these patients depend on immunosuppressive drugs, and the long-term effects of such treatments are not yet clear. There is therefore much research on other ways and means of preventing rejection. Structures that suffer from wear or tear can be replaced by mechanical components. For a long time it has been possible to provide artificial teeth, and more recently methods have been developed for the replacement of hip joints and heart valves with mechanical components that are not rejected. In the future there will be further advances and more elaborate and expensive replacement techniques will be developed. This raises important issues about the costs of medical care for the aged, which are discussed at the end of the chapter.

There is much ill-informed discussion about the replacement of components of the central nervous system. For example, could there be eye transplants in the future? The problem here is on a completely different level from, say, a heart transplant. There are vast numbers of axons from the rods and cones of the retina that extend to the tectum of the brain, and vision depends on a perfect match between input of the signal and its transmission to a specific site in the tectum. The cutting of the optic nerve destroys every connection, and it is surely unreasonable to envisage repair of these. Similar arguments apply to the repair by transplantation of defective parts of the brain. This does not mean that treatment of brain diseases will be impossible in the future, but transplantation is very unlikely to become a viable procedure.

MODULATION OF LONGEVITY

There is much popular interest in treatments or life styles that prolong lifespan. The simplest and most effective is calorie deprivation in mice and rats. This work was pioneered by McCay et al. (1939), and in every subsequent study the effect has been confirmed. Moreover, the comparison of animals fed ad libitum and calorie-restricted animals has become the focus of many experimental studies (see Chapter 7). Starvation-induced increase in lifespan may be an evolutionary adaptation (Holliday 1989b), and this raises the question of the effect of reduced food intake in other mammalian species. We do not yet know the answer, although it is possible that longitudinal studies in humans, or other epidemiological studies, may provide it for our species. It has also been known for some time that removal of the pituitary gland of rats (hypophysectomy) significantly increases lifespan, but no explanation for this is yet available (Everitt et al. 1968, 1980; Everitt & Meites 1989). It is significant that there are no recent reports of antioxidants increasing animal lifespan. Early results were not confirmed when appropriate long-lived mice were used (see Schneider & Reed 1985). In view of the great current interest in the role of oxygen free radicals in ageing, it would be surprising if more recent longevity experiments had not been completed. The absence of any positive results suggests that results may have been negative and remain unpublished.

There is more uncertainty about treatments that reduce longevity of mammals, because one has to separate toxic effects from a genuine acceleration of ageing. In the many studies using ionising radiation, the evidence strongly suggests that the treated animals have the same age-related changes as untreated controls later in life (Lindop & Rotblat 1961). Obviously, human life style can influence longevity; overeating and lack of exercise are likely to result in late-onset diabetes, with all its side effects, and vascular disease. In a specific test of the protein-error theory, young CBA inbred mice were treated with an amino acid analogue in drinking water for just four weeks. Thereafter, they were perfectly normal, but some of the treatment regimes reduced lifespan. This effect cannot be attributed to toxicity per se, but must be due to a long-delayed consequence of an initial effect on proteins. The treatments that increase or decrease the lifespan of rodents are listed in Table 9.1.

Other studies have been carried out with cultured human fibroblasts. There are several treatments that either increase or reduce lifespan (Table 9.1). The

Table 9.1 *Some treatments that increase or decrease the lifespan of mice or rats, and cultured human fibroblasts (see also Note 9.1)*

	Treatment	Effect on lifespan	References
Mice or rats	Calorie restriction	Increase	Reviewed by Schneider & Reed (1985); Holehan & Merry (1986)
	Hypophysectomy	Increase	Everitt et al. (1980); Everitt & Meites (1989)
	Ionising radiation	Decrease	Neary (1960); Lindop & Rotblatt (1961)
	pFPA treatment[a]	Decrease	Holliday & Stevens (1978)
Human fibroblasts (PDs)	Hydrocortisone	Increase	Macieira-Coelho (1966); Cristofalo (1974); Cristofalo & Kabakjian (1975)
	Ionising radiation[b]	Increase	Kano & Little (1985); Croute et al. (1986); Holliday (1991b)
	Antisense to p53 and Rb	Increase	Hara et al. (1991)
	Carnosine	Increase	McFarland & Holliday (1994)
	Fluorouracil	Decrease	Holliday & Tarrant (1972)
	Paromomycin	Decrease	Holliday & Rattan (1984)
	5-azacytidine or deoxyazacytidine	Decrease	Fairweather et al. (1987); Holliday (1986a)
	Incubation at 40°[c]	Decrease	Thompson & Holliday (1973)

[a]3-4-week or 1-year-old CBA mice were treated for four weeks with 4×10^{-4} *M* para-fluorophenylalanine (pFPA) in their drinking water; 2×10^{-3} *M* pFPA had a non-significant effect.
[b]Moderate doses increase lifespan; higher doses decrease it (see also Macieira-Coelho et al., 1977).
[c]Incubation at 34° C had no effect on lifespan.

mechanisms of life extension are far from clear, and few biochemical studies have yet been carried out. The tumour suppressor genes p53 and Rb have a role in preserving the phenotype of normal cells; their partial inhibition with oligonucleotide antisense RNA may push the cells towards an immortal phenotype, but this effect is transient. Carnosine is a naturally-occurring dipeptide that is present at a high levels in tissues such as muscle and brain. It is an antioxidant and may prevent non-enzymic glycation of proteins (see Note 9.2). Carnosine has remarkable effects on cultured human fibroblasts *in vitro*. The cells preserve a much more normal morphology as they approach the end of their proliferation. At high concentrations, such as 30 or 50 m*M* in growth medium, few of the normal signs of senescence are seen (see Chapter 5 and Fig. 5.2), but when these late passage cells are transferred to normal medium they show the typical senescent phenotype after a few days. Also, transfer of senescent cells in normal medium to medium containing carnosine can rejuvenate them (McFarland & Holliday 1994). It is possible it has an important role in preserving normal cellular homeostasis. It is notable that ionising radiation, which has a life-shortening effect on animals, appears to significantly extend the Hayflick limit to fibroblast growth. This may be due to the induction of longer-lived clones that take over the late passage cultures. Environmental treatments that either increase or decrease the lifespan of cultured human cells provide some evidence against the loss-of-telomere theory of cellular ageing. The telomere clock should be dependent on cell division, and additional assumptions have to be made to account for environmental effects. For instance, the size of the terminal overhang (see Note 5.5, Fig. 5.8) might vary under different conditions of growth.

Treatments that increase lifespan always arouse interest, and those that reduce it are usually dismissed as examples of delayed toxicity. In fact, both effects are important in the study of ageing, although it must be admitted that so far they have not illuminated our understanding of underlying biochemical processes.

SINGLE OR MULTIPLE CAUSES?

A remarkable feature of ageing is that various organs or structures have evolved to 'last a lifetime'. Teeth wear out, atherosclerosis develops in arteries, the valves of the heart become calcified, tumours develop, brain cells are lost, joints become arthritic, and so on and so on. There is a degree of synchrony in all these senescent changes, although as we all know, any one tissue or organ system can degenerate in advance of others. It is then usually labelled as a disease, which is commonly thought to be distinct from 'natural ageing'. The gradual loss of function of different organs means in one sense that there are multiple causes of ageing, but it is still possible that one type of molecular change is mainly responsible for all senescent changes. In Chapters 3 and 4, maintenance mechanisms and theories of ageing were discussed. It is clear that all maintenance mechanisms are responsible for the preservation of the adult, and it was proposed that their eventual failure gives rise to the senescent or aged phenotype. Many of the theories of ageing are related to the failure of a specific maintenance mechanism (see Table 4.3), and it was proposed that a more global view of ageing would encompass aspects of all theories. This in turn implies that there are multiple causes of ageing at the cellular and molecular level, as well as at the level of tissue and organ. These same multiple causes would operate both in long-lived mammals, such as humans, and short-lived ones, such as rodents. The fundamental difference between the two would be the amount of metabolic resources invested in maintenance, as discussed in Chapter 7. Obviously, if the enzyme systems that repair damage in DNA are more efficient in man than mouse, then the integrity of the DNA in the former will be preserved longer than in the latter. Not so

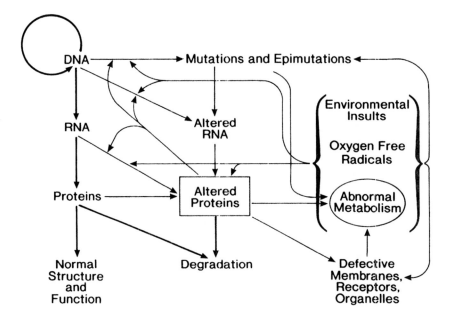

Figure 9.1. Interrelationship between the normal pathways of information transfer, and the various defects that can arise in cells. In normal cells, defects may occur, but the levels of feedback, or deleterious interactions, do not result in a continual increase. Therefore homeostasis is maintained. It is suggested that there is a significant probability of losing this steady state, and this will eventually lead to a breakdown in the normal metabolic controls and cell death (see text). The heavy arrows indicate normal pathways of information transfer and removal of defective proteins. The thin arrows indicate all abnormal events.

obvious are the ways in which other maintenance or homeostatic mechanisms may act, for example, in setting the rate of collagen cross-linking, or the loss of transparency in the lens. Nevertheless, alterations to proteins could certainly be modulated by maintenance, for example, in the defences against damaging oxygen free radicals, the level of soluble sugars, or other compounds that can interact with proteins. In some cases the evolved amino acid sequences may be related to stability and function (Robinson et al. 1970).

Although maintenance mechanisms involve quite different molecular and cellular systems, there is nevertheless much overlap and interaction. In Chapter 3, I referred briefly to cellular homeostasis, but further elaboration is necessary if we are to understand why cells do not last indefinitely. Possible interactions between essential molecules are illustrated in Figure 9.1. There is a given level of damage in DNA, and failure of repair or error-prone repair produces mutations, epimutations and chromosome abnormalities. Mutations can produce mRNA and proteins with the wrong sequences of bases and amino acids. Errors in the synthesis of RNA are known to be about 10^4 times higher than in DNA, and a similar or higher level of errors occurs in the translation of mRNA into polypeptides. These altered molecules may have feedback effects on pathways of RNA, protein and DNA synthesis. Proteins are also altered in a variety of ways, which were discussed in Chapter 4. These too can disrupt or reduce the accuracy of macromolecular synthesis. There are active proteolytic mechanisms that degrade altered proteins, but they too may be subject to a reduction in efficiency, which would result in the accumulation of altered proteins. Once this starts to happen, the cell is no longer in a steady state. Altered proteins also produce abnormalities in metabolism, which will intrude on the cellular mechanisms regulating the normal interactions between macromolecules. Amongst other things, this may affect epigenetic controls, leading to the ectopic expression of genes that would otherwise be silent. Membranes and organelles would also be directly affected by the presence of abnormal molecules, and membranes may harbour defective receptors. Thus the traffic in small molecules and metabolites in and out of cells may be affected as well as the signalling between cells. Oxygen free radicals are primarily generated by respiration in mitochondria. Any imperfections in the mitochondrial membrane is likely to lead to the leakage of free radicals, and therefore to damage to other cell components. Lipid peroxidation may occur and this may bring about the accumulation of lipofuscin. As well as free radicals, there may be environmental insults, such as toxins in food or their by-products, which damage macromolecules or disturb normal metabolism. It is, in fact, all too easy to see how things might go wrong in normal cells, and how once things have gone wrong, there can only be an ever-increasing level of defects leading to cell death. This is not an error catastrophe in protein synthesis, but a loss of homeostasis, or a cell catastrophe *in toto*. In some situations when cellular components are damaged, by external agents or intrinsic events, the suicide mechanism known as apoptosis is switched on. This may be advantageous for the organism if that cell can be replaced by a normal one, but there is little evidence that apoptosis is related to the senescence of cells *in vitro,* or in ageing tissues.

Thus, although there are many possible kinds of defects in cells and many maintenance and regulatory mechanisms that remove such defects, there is also an enormous number of possible interactions between metabolites and macromolecules. Unrecognised or unrepaired defects can therefore influence or alter a wide variety of other cellular components. The upshot is that many possible causes may have one end result, namely, the death of the cell. At first sight, the existence of inherited diseases that accelerate ageing (see Chapter 8) suggests that there could be a primary or single cause of ageing, but it is not difficult to see how a single molecular defect could precipitate cellular abnormalities or death in many tissues. For example, in Figure 9.1, a defect in an RNA polymerase, a protease or a component of a mitochondrial membrane could initiate a chain of other events and bring about premature cell death. Another possibility is a defect or abnormality in normal protein phosphorylation, which is essential for normal cellular regulation and signal transduction. The defect could be in a protein kinase, or in the protein that is normally phosphorylated. These possibilities may make it easier to understand how a single genetic defect, such as the recessive mutation that gives rise to Werner's syndrome, can have pleiotropic effects on the individual. Certainly it is very important to identify the actual gene affected in Werner's syndrome or progeria, and this will advance our knowledge of ageing, but it is doubtful if it will uncover a single specific cause of ageing. Similarly, if the premature ageing seen in Down's syndrome is due to an extra gene coding for superoxide dismutase (SOD), then we would conclude that too much of the enzyme may lead to a high level of hydrogen peroxide and hydroxyl radicals (see Fig. 3.1). This certainly would implicate oxygen free radical damage in the ageing process, but would not allow one to assume that an excess of the hydroxyl radical is the cause of ageing.

In his review of the inherited conditions that, in one way or another, affect degenerative changes associated with ageing, Martin (1978) concluded that very many genes are involved. On the other hand Cutler (1975) argued that increased longevity evolved very rapidly in the higher primates, and therefore few genes are involved. Similarly, the search for 'gerontogenes' in experimental organisms implies that specific genes strongly influence or control ageing.

Perhaps both viewpoints can be defended if we consider the question Hayflick (1987) raised: 'Why do we live as long as we do?' We survive because structure and function are maintained by a variety of mechanisms. These mechanisms depend on enzymes, proteins and many cell types, and a very large number of genes are required to specify all this information. Thus, ongoing survival depends on many genes, but the imperfections of maintenance in its various forms also implicate many genes. Nevertheless, the complex interactions between various molecular and cellular mechanisms mean that a single defect can have many deleterious consequences. Just as a single spanner thrown into a complicated machine can terminate function, so the loss of a single gene can have the pleiotropic effects that are seen in premature ageing (see also Note 9.3). It is correct to refer to 'multifactorial ageing' as Olson (1987) did, because so many changes occur in the cells and tissues of the body. Yet it is the interactions between molecular processes, or the molecular basis of cell homeostasis, that may be the key determinant in setting the efficiency, and the rates at which various maintenance mechanisms eventually fail.

MOLECULAR BIOLOGY OF AGEING

In the Preface I stated that this book would attempt to provide a broad overview of the underlying mechanisms of ageing, and did not claim that we understand the fine details. This information will come from future studies and especially the molecular biology of ageing. Although molecular studies have not been emphasised in the text, the understanding of many of the essential processes reviewed in Chapter 3 have already been greatly advanced by molecular methods.

One approach that will receive much attention in the future is the identification of genes that have strong effects on the rate of ageing. Progress is already being made in the search and cloning of important human genes, such as that responsible for Werner's syndrome. The identification of the gene or genes responsible for the phenotype of SAM mice (Chapter 6) will also be a very important advance. Other genes that are likely to be important include DNA repair genes, and those coding for enzymes that are important in the defence against oxygen free radicals. Indeed, there is now a large amount of research in progress on the damage caused by free radicals. This includes molecular studies, for example, the detection of deletions in mitochondrial DNA, which are presumably caused by free radical attack.

If one considers all the maintenance mechanisms reviewed in Chapter 3, it is clear that a wide range of cellular, biochemical, molecular and genetic procedures will be brought to bear in their further study. This in turn will throw much light on the reasons for failure of maintenance and the age-related changes associated with senescence. It will be important to understand the interrelationships of all these future findings, so that the underlying causes of different age-related diseases can be revealed.

In such studies, it is very likely that comparative studies using different mammalian species will become very important. The fact that very similar changes occur at very different rates in short- and long-lived species may provide the key to understanding the underlying molecular processes. For example, if we knew why cross-linking of collagen occurs more rapidly in rats than in humans, we would probably also understand the molecular mechanism of cross-linking. One formidable problem, of course, is that the experimental animals chosen for study are not necessarily appropriate models for human ageing. This is very clearly shown in studies of cancer (Chapter 8). Research is very commonly carried out with rodent cells, but it is already very well known that these are far more susceptible to neoplastic transformation than human cells. Much work on oncogenes, tumour suppressor genes and so on, which has been done with rodent cells, needs to be extended to the study of human cells. Unfortunately, such approaches are all too often put aside, because human cells are thought to be too 'difficult' to work with. Yet to understand why it takes 20 or more years for tumours to appear in human tissues requires not only the identification of underlying molecular events, such as mutations or chromosomal changes, but also the *defences* against tumour progression. It is obvious that these defences, or the maintenance of cell integrity, are much more effective in humans than in short-lived species. This highlights some of the problems in using transgenic mice to study genes known to be important in humans. It is obvious that in the future there will be more and more transgenic mice lacking genes, or with mutant genes, that are known to be important in our species. These studies will be extended to genes that have important effects on ageing. Nevertheless, care is necessary in extrapolating from mouse to man, and already there are some surprises. It is reported that mice without HPRT enzyme activity are normal, whereas humans are severely disabled (Lesch–Nyhan syndrome), and there are major differences in the roles of oncogenes and tumour suppressor genes in mouse and man. This does not mean that transgenic animals are not essential in ongoing research, but for understanding the molecular basis of age-related changes, the mouse may be a poor model for humans, and experiments with larger longer-lived animals are difficult, expensive and may take a long time to complete.

One very important experimental approach that is already being attempted involves the cloning of the enzyme telomerase. If dividing somatic cells have limited lifespan because telomeric DNA is continually lost, then the production of cells with an active gene for telomerase should lead to their immortalisation. This would be a major advance and, no doubt, be accompanied by much publicity. It has already been suggested that transgenic mice with active telomerase in all their cells would live forever. This is naïve in the extreme, since all post-mitotic cells would be unaffected.

BIOMEDICAL RESEARCH AND AGE-RELATED DISEASE

In ending this book, I wish to emphasise the importance of fundamental research on ageing. In the Preface and in the book as a whole I have claimed that we now have an understanding of ageing at the biological level. I have also said that the fine details of ageing at the cellular and molecular levels will be understood in the future. Why is this research so important? The answer comes from a consideration of age-related diseases, some of which were briefly discussed in Chapter 8. In countries with good health care, the proportion of elderly people is continually increasing, and therefore the number of individuals with one or another age-related disease is also increasing. The medical treatment of many of these individuals is very expensive, for example, by vascular surgery or organ transplantation. Moreover, the law of diminishing returns applies to medical treatment of the elderly, since all too often treatment of one disease is followed by the appearance of another. The cost of medical care for individuals over 65 is five or six times higher than for younger members of the community, and about 50% of individuals over 65 have some physical disability. In developed nations the costs of health care as a percentage of gross national product doubled between 1960 and 1986 (WHO 1991 statistics). Much of this increase is due to the increasing costs of health care for the aged. The costs in real terms will double again early in the 21st century, unless of course means are found to cut this enormous expense.

Instead of treating each disease as it arises, or worse, devising even more expensive treatments, it is essential that there be a drastic reappraisal of medical practice and biomedical research, preferably now, but otherwise in the next century. This viewpoint is not at all new. The geriatrician Edward Stieglitz (1942) wrote in an article on 'The social urgency of research on ageing':

> The shifting age distribution of the population with its increasing proportion of those in the older age groups has introduced innumerable problems of the most practical and significant character. The situation is without precedent. The millions of elderly are here. There will be more. Many are well and capable of continued productive and creative effort if given opportunity to work within their capacities. Others, and there are many, are prematurely disabled by the insidious chronic and progressive disorders so frequent during the senescent period and become heavy burdens upon the family and society by reason of the long course of their disablement. . . . Gerontology, the science of ageing, is divided into three major categories: (a) the problems of the biology of senescence, (b) the clinical problems of ageing man, and (c) the socio-economic problems of ageing mankind. These three categories are intimately and inseparably related; progress in one field is dependent upon progress in the others and *vice versa.* . . . Clinical medicine, in its youngest speciality, geriatrics, can do much to solve the most distressing problems in the social field. To name but two potential contributions of vital importance: Personal preventive medicine may greatly reduce the toll of premature disability and bettered diagnostic methods can more clearly define the limitations which go with normal ageing. Yet even more intimate is the dependence of clinical medicine upon advances in the fundamental sciences in elucidating just what ageing is, what it does, what retards or accelerates it and why. It can not be over-emphasized that *the more we know about the biologic mechanisms of the ageing processes, the more effectively can clinical medicine treat the ageing and the aged.* (pp. 905–6; italics added)

Fifty years later there are very few signs that those who research age-related diseases, or those who have the responsibility of treating such diseases, are at all convinced that fundamental research on ageing itself is of prime importance (see also Holliday 1984d).

There is now an enormous amount of research on age-related disease, and this research has three basic aims: first, to improve treatment of the disease in question; second, to prevent or delay the onset of the disease, and third, to understand the aetiology or cause of the disease. These three approaches are laudable, and cannot be criticised. The problem resides in the plain fact that almost all those working on age-related diseases do not believe they are in the field of gerontology. It is very unfortunate that medical opinion insists on making a distinction between a specific disease or pathology and so-called natural ageing. This is even more surprising when everyone knows that the death certificate provides one or more causes of death, which are pathological conditions: it is not permissible to attribute death to natural ageing. So we have a situation where there are laboratories and institutes that study cancer, that study vascular disease, that study Alzheimer's disease, that study osteoporosis, that study late-onset diabetes, and so on and so on. There are specific journals, as well as conferences and conferences, devoted to each disease. Most of this effort cannot in itself be faulted, but what is wrong is the failure of the various research scientists and specialists to communicate with those working on some other age-related disease. In general, diabetes specialists would not interrelate with, or think they have anything in common with, cancer specialists. Experts on Alzheimer's disease do not communicate with those working on atherosclerosis. There are some pathologists who take a broad view of ageing and age-related disease, but they appear to be a small minority. The majority of biomedical scientists working on one or another disease do not consider themselves to be in the realm of gerontology, and the huge literature that documents all current knowledge of these diseases is not considered to be part of this field.

This is what needs to be changed if we are to better understand the causes of age-related disease, and to devise methods for their prevention or delay of onset. What is required is fundamental research on ageing per se, particularly at the cellular and molecular levels, and far more interaction amongst the various biomedical fields of research to which I have referred. The overall aim of ageing research is to make treatments, especially expensive treatments, much less common by preventing or delaying the onset of the disease. In the long run this will save enormous health care costs. These saved resources could then be put to other uses, for example, in gaining more information about inherited diseases, or providing better treatment for younger members of the community. As well as that, the delay or prevention of age-related disease will greatly increase the quality of life of innumerable elderly people. They will still age, but they need not so often be afflicted with pain or loss of faculties. There would continue to be an increase in the expectation of life, but it would be a relatively modest increase, perhaps comparable to what has occurred over the past several decades.

This provides not only a justification, but emphasises the real need to have more and better research on ageing. It is all too often assumed that those who are interested in ageing are directing their research towards the goal of extending human lifespan. I have tried to explain in this book why this aim – at least in extreme form – is an illusion. Yes, we can delay the features of ageing brought about by disease; yes, we may be able to influence ageing by changes in diet or life style, but a search for the elixir of life is as fanciful as that for the transmutation of other metals into gold.

Notes

1.1 Benjamin Gompertz was an English 19th-century actuary who concluded that the 'The number of living corresponding to ages increasing in arithmetical progression, decreased in geometrical progression'. In simple terms, this means that a plot of age on a linear scale against log of the mortality rate gives a straight line. The age-specific mortality rate is the fraction of survivors that dies in the next time interval. Various forms of the basic Gompertz equation have been discussed in the gerontological literature in relation to survival curves (see Finch 1990). A good fit is usually obtained for experimental animals, such as mice or rats, in a uniform environment, and for humans with good health care, but any Gompertz-derived function is necessarily an approximation that will fit some sets of data better than others. Recently, it has been shown that the mortality rate of some insects is very different from the Gompertzian prediction (Carey et al. 1992; Curtsinger et al. 1992).

1.2 In very early cell culture experiments, Alexis Carrel grew chick fibroblasts for long periods of time. He claimed that they survived in continual culture for about 30 years, which is longer than the lifespan of the donor animal. His procedures have been strongly criticised (see Witkowski 1980), and more recent work with primary chick cultures shows that they grow for about 30–40 population doublings. When methods for growing mammalian cells were well defined, it was found that tumour cells, such as the HeLa cell line, would grow indefinitely. Mouse cells, such as the L cell line, would also grow indefinitely, but this is derived by transformation from a normal diploid cell population. The distinction between diploid cells with defined lifespan and heteroploid cells that are 'immortalised' was first made clear by Hayflick (1965). He recognised the equivalence of transplantable heteroploid tumour cells and permanent lines in culture, and the equivalence of normal diploid somatic cells and primary cells in culture (see also Witkowski 1987, for a historical account.)

2.1 Comparison with DNA immediately comes to mind. The DNA molecule is a duplex so there are two complementary copies of genetic information. Most DNA damage affects only one strand, so the other strand can be used as a template to repair the damage (see Chapter 3). For foolproof repair it would be desirable to have three copies, so that any abnormality can be recognised in comparison with two normal copies. Obviously such a structure would be difficult to replicate. Instead, when a DNA molecule is damaged in both strands (as in a double-strand break, or a cross-link between the strands), another homologous molecule in the cell can be used as the correct template to repair the damage by a recombination mechanism. This may have been a major reason for the evolution of diploid organisms from haploid ones.

2.2 There is a problem in the interpretation of transplantation experiments if the populations used initially consist of stem cells and non-stem cells, as is the case with haematopoietic cells. In the initial transfer the stem cells are in minority, and the proportion is very likely to decline with each sequential transfer. Thus, the stem cells can be lost by dilution, and then the remaining cells will eventually die out, even if stem cells are potentially immortal. This is discussed further in Chapter 5, especially in relation to the commitment theory of cellular ageing.

2.3 The hair follicle is a complex structure containing many types of cells. Melanoblasts are produced by the neural crest during development and migrate over the surface of the animal. They give rise to melanocytes, which produce the pigment granules in coloured hair, and hair without these granules is white. The white-spotting gene, which is known in many mammalian species, severely reduces the number of melanoblasts, so when they migrate, large areas of the underside of the animal do not have any melanocytes and remain white. Radiation, which kills off dividing melanocytes, can convert coloured mice into white ones (provided part of the animal is shielded to preserve bone marrow cells). At higher doses, hair follicles become non-functional and hair is lost.

3.1 Proof-reading mechanisms to increase the accuracy of protein synthesis were proposed independently by Hopfield (1974) and Ninio (1975). In Hopfield's 'kinetic proofreading', there are one or more intermediate reaction steps in which the enzyme–substrate complex is driven into a high-energy transitional state from which there is an increased probability of dissociation of an incorrect complex. This could occur, for example, in the charging of specific tRNAs with the wrong amino acid. The accuracy can be increased by having additional steps of proof-reading, but more energy is then required, so as in all accurate synthesis in biological systems, there is an appropriate optimum. In Ninio's scheme each step of synthesis is subjected to a time delay, during which the incorrect substrate enzyme complex dissociates more quickly than the correct one. This causes a slowing of the rate of correct synthesis, in accord with the general principle that rapid synthesis leads to more errors and slower synthesis to greater accuracy. Again, an optimum is reached. Mutants of ribosomal proteins are known in *E. coli* that increase the accuracy of translation, but slow down the rate. For a full discussion of proof-reading, including the experimental evidence, see Kirkwood, Rosenberger & Galas (1986) and Jakubowski & Goldman (1992).

3.2 The existence of many introns in structural genes demands a mechanism that ensures accuracy in splicing, that is, the correct joining of adjacent exons without the addition and deletion of RNA bases. Since some genes have 50 or more introns, clearly a special mechanism is required. The consensus splicing sequence does not provide sufficient accuracy, and in any case the same sequences may lie within individual exons or introns that are not splice sites. The problem is compounded by the phenomenon of alternative splicing, in which given sets of exons in a structural gene produce one spliced together to form alternative mRNA, and a family of related protein gene products (see Breitbart et al. 1987; Smith et al. 1989). It is unlikely there could be specific proteins that mediate each splicing step. A more attractive possibility proposed by Murray & Holliday (1979a,b) is based on RNA recombination. In DNA recombination, accuracy is achieved in the breakage and rejoining of molecules by complementary base pairing (Holliday 1964). The same principle can be applied to splicing if small RNA molecules exist that hybridise to adjacent exon

Figure for Note 3.2. Accurate splicing of adjacent exons may be mediated by short RNA molecules, which exactly hybridise to the terminal exon sequences. Subsequently the intron is removed and exon termini are ligated. Other splicer RNA molecules could mediate alternative splicing, in which exon 1 may be joined to exon 3, and so on.

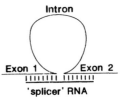

sequences (as shown in Fig. to Note 3.2), or alternatively, to the ends of intron sequences. This brings the ends of the exons and introns together, and an enzyme complex (such as SnRNP) could cut and rejoin all such structures, whatever the exact base sequences. The hypothesis predicts the existence of many small 'splicer RNAs' with appropriate specificity. Alternative splicing would depend on the regulation of the transcription of such molecules.

3.3 The accuracy of protein synthesis is discussed more fully in Chapter 4, but it is important to realise that many types of error can occur. One might ask, for example, what the probability is of substituting one amino acid at a given position for another similar amino acid, such as isoleucine for valine; but there are 19 different possible incorrect substitutions at each position. For a polypeptide of 200 amino acids, this means there are $200 \times 19 = 3800$ possible types of error. Obviously, some are much more likely to occur than others. An error level of 10^{-4} could refer to an individual type of substitution, or the sum of many substitutions. Loftfield (1963) was a pioneer in first attempting to estimate the accuracy of protein synthesis. Subsequent information about accuracy is very sparse, particularly in mammalian cells (see Kirkwood et al. 1984, 1986). Not much more information has been gained in subsequent years.

3.4 The adducts excised from DNA by repair enzymes (especially DNA glycosylases) include thymine glycol, thymidine glycol, 8-hydroxyguanine, 8-hydroxy deoxyguanosine and 5-hydroxymethyl uracil. Their excretion in urine is a demonstration of the defence against continuous damage to DNA that occurs, but it is hard to quantitate the actual number of altered bases per cell per day, and the efficiency of their removal. It has been estimated that there are about 10 000 oxidative hits to human DNA/cell/day and perhaps ten times more in rat DNA (Ames et al. 1993), but these estimates may be too high (Lindahl 1993). With regard to efficiency in defence, the important parameter is the number of altered bases in DNA that are *not* removed by repair (see also Chapters 4 and 7).

3.5 In bacteria, restriction and modification enzymes recognise short specific sequences in DNA, and the former can distinguish between unmodified and modified DNA. The idea that modification of DNA in higher organisms might provide specific signals, or epigenetic information, was proposed independently by Riggs (1975) and Holliday & Pugh (1975). In both cases, proteins were invoked that could distinguish between modified and unmodified DNA substrates, and it was noted that the pattern of modification can be stably inherited. The commonest form of modification in higher organisms is 5-methyl cytosine (5-mC), although it is now known that 6-methyl adenine also exists in chromosomal DNA (Kay et al. 1994). 5-mC occurs primarily in CpG doublets in DNA, but this doublet does not provide specificity. Therefore if DNA methylation is important in the control of gene expression, then nearby base sequences would provide the required specificity for the binding of regulatory proteins, which can distinguish between a modified and an unmodified sequence. In fact, there is now much evidence that methylated sequences are frequently associated with lack of transcription, or the silencing of genes (reviewed in Adams 1990; Holliday, Monk & Pugh 1990; Jost & Saluz 1993). Since 3–5% of cytosines are methylated in mammalian DNA and CpG itself is not a very specific signal, it is likely that only a subset of 5-mC residues is important in gene regulation, and the rest have some other role, such as a structural function in the organisation of chromatin or chromosomes.

There is now much evidence that the presence or absence of methylation is inherited in biological systems, in many cases with considerable accuracy (see Holliday 1993). The biochemical basis of this is not fully understood, because purified DNA methylases will act on non-methylated substrates *in vitro,* although they prefer a hemimethylated substrate. Accurate maintenance *in vivo* may require the participation of one or more other proteins. Since the pattern of methylation is highly specific for individual tissues (see Doerfler 1993), there must be mechanisms that introduce methylation at particular sites, and that also remove it. These mechanisms are not as yet understood.

3.6 At the 5′ end of many genes there are CpG islands, which comprise about 1% of mammalian DNA (reviewed by Gardiner-Garden & Frommer 1987). The rest of the DNA is CpG depleted, but in islands the number of CpG doublets is roughly what is expected from the overall base composition. There are about 30 000 CpG islands in the mammalian genome, which is thought to correspond broadly with the number of genes; indeed CpG islands are markers for structural genes. Islands are not methylated and they are usually associated with active genes, whereas CpGs in non-island DNA are very often methylated. An important exception is that CpG islands on the inactive X chromosome are methylated. It is often maintained that CpG is depleted in bulk DNA because 5-mC is spontaneously deaminated to thymine, converting a CpG doublet to TpG and its complement CpA. However, this view takes no account of the actual functions of 5-mC residues (see Holliday & Grigg 1993).

3.7 The theoretical physicist Erwin Schrodinger wrote an influential book, *What Is Life?* (Schrodinger 1944), before the advent of molecular biology. The physical laws of thermodynamics are statistical because they deal with the property of large numbers of molecules. Schrodinger realised that the information about genes from classical genetic studies was not compatible with statistical laws, because genes are single entities that are very stable. Not only do they mutate at a low rate, but the new form of the gene is also stable. The apparent order of the genetic system appeared to be contrary to what would be expected from the random motion of individual molecules. He therefore suggested that there might be new laws to be discovered that would explain the particular properties of genes. In fact, the unique properties of genes are explained by the stable structure of DNA and the enzymes that replicate it. There is nothing contrary to the laws of physics or chemistry in the behaviour of the genetic material. Nevertheless, Schrodinger was correct in his realisation that living material defies the statistical laws of thermodynamics, and creates order from disorder by continually taking energy from the environment. As he put it: organisms feed on negative entropy.

4.1 When DNA is damaged, the vast majority of lesions are accurately repaired and no mutations arise. Mutations arise in two circumstances. First, an abnormal base produced by a mutagen, such as an alkylating agent, may form a mispair during replication, and the same can be true if base analogues are incorporated into DNA. It is probable, however, that abnormal bases are usually excised before they have the opportunity to form mispairs at the replication fork, or the mispair is recognised and corrected. Second, if a lesion cannot be accurately repaired, then a second line of defence is 'error-prone' repair, which may generate new mutations. SOS repair in bacteria is error prone, and it comes into play when the normal DNA polymerase cannot replicate past a lesion in DNA. Under these circumstances error-prone repair allows replication to proceed, but with the likely insertion of an incorrect base opposite the lesion (see Friedberg 1985).

4.2 Because enormous resources are invested in human health care, a very large number of inherited diseases have been identified. These have been catalogued over many years in McKusick's *Mendelian Inheritance in Man,* which has now reached its ninth edition (1990). The number of gene loci identified in this edition is 2656, with a further 2281 loci that are less rigorously validated. (In the first edition the corresponding numbers were 574 and 913.) Of the fully confirmed loci 1864 are autosomal dominant, 631 are autosomal recessive and 161 are X linked. One might expect the majority of mutations that knock out the function of a gene to be recessive, but they are only expressed when inherited from both parents. Dominant and codominant mutations are expressed directly in offspring, hence the larger number identified. Nevertheless, the number of dominant mutations is surprising, since one might expect many of them to be defects in the regulation of genes rather than in the coding regions of genes. It is possible that a substantial number of dominant and codominant mutations are in genes coding for polymeric proteins, in which one or more altered subunits affects function.

4.3 The remark about the age of election of Privy Councillors is attributed to John Maynard-Smith in response to a presentation by Philip Burch. Burch (1968) maintained that age-related pathologies, as well as such features as whitening of the hair or dental caries, could be explained by sequential mutations, and in each case it was possible to derive a particular formula for the origin of the defects. One could say that the career of a Privy Councillor also depends on sequential events, and only when all these have been achieved is election recommended. The likelihood of achievement obviously increases with age. The same question of interpretation arises in discussions of the age-related incidence of cancer. It is agreed that carcinomas arise from several sequential events, but it is disputed whether cancer is related to ageing (see discussion in Chapter 8).

4.4 Orgel (1963) originally suggested that errors in protein syntheses would continue to increase by a feedback mechanism, in the absence of selection. Later, he reconsidered the kinetics of error increase in a simpler model, and drew the conclusion that there are two outcomes depending on the extent of error feedback (Orgel 1970). This was defined by the parameter α. If $\alpha > 1$, then errors continually increase, as he previously proposed, but if $\alpha < 1$, then the errors reach a steady state. Experiments in *E. coli* by Edelmann and Gallant (1977b) suggested that α is about 0.8, which is not far removed from the critical level of 1.0.

A more detailed theory was proposed by Hoffman (1974), who concluded that protein synthesis was very stable in present-day organisms, and the probability of error feedback and an error catastrophe was negligible. However, in Hoffman's model, protein molecules containing errors had almost no activity, which is clearly untrue (Kirkwood & Holliday 1975a). A new parameter R was introduced, which is the residual activity of an enzyme (or 'adaptor') containing an error. From this was developed the Hoffman–Kirkwood–Holliday (HKH) model, which is much more biologically realistic than Hoffman's earli er one (Kirkwood 1980). Starting with a hypothetical error-free system, subsequent generations of proteins have increased errors, but soon reach a threshold, which is a metastable state. There is a given probability that chance fluctuations in accuracy, or critical errors, could drive the cell beyond the threshold, and then errors would increase exponentially to a lethal error catastrophe. It is obvious that organisms have evolved optimum values for accuracy in the synthesis of DNA, RNA and proteins. For enzymes, the important parameters are catalytic activity, substrate specificity, and stability (see Kirkwood & Holliday 1986a). There are complex interrelationships between these, and in terms of accuracy, it is difficult to know how far removed the protein synthetic apparatus is from the point where error feedback becomes critical and stability may be lost. (For a recent discussion see Kowald & Kirkwood 1993.)

4.5 Cellular proteins can be separated in two-dimensional gels, where separation depends on molecular weight in one dimension, and charge in a second dimension. If random errors are introduced, then size is not affected, but a proportion of molecules with altered charge will be produced. These produce minor 'stutter spots' on each side of the main protein spot. Only a minority of amino acid substitutions change charge, and probably at least 5–10% of molecules would have to have at least one additional negative or positive charge to produce a visible stutter spot. This means that the method can detect substantial numbers of errors, but not necessarily significant increases over the background level.

4.6 Harley et al. (1980) treated human cells with histidinol, which is an analogue of histidine. This blocks the normal charging of histidyl tRNA, and as a result other amino acids are incorporated at 'hungry' histidine codons. Young and old human fibroblasts produced roughly the same level of stutter spots, and from this they concluded that the *spontaneous* levels of errors in these cells are the same. They use a theoretical argument that is based on some unjustified assumptions. One is that the levels of histidyl tRNA in old and young cells are the same, and another is that ribosomes from senescent cells, if ambiguous, would have increased affinity for non-cognate amino acids, but unaltered affinity for cognate amino acids. The truth is that the error level in protein synthesis in young and senescent cells has not yet been determined.

4.7 The first step in non-enzymic glycosylation is the reaction between the aldehyde group of a reducing sugar and the free amino group of a protein. This is often the ε-amino group of lysine, but it can also be the terminal amino group of a polypeptide chain. The first product is a labile Schiff base, which undergoes a rearrangement to form a more stable Amadori product. These are the early glycation products and the reactions are reversible:

$$R_1\text{---CHOH---C}{=}O + NH_2\text{---}R_2 \Leftrightarrow R_1\text{---CHOH---}\overset{\overset{\displaystyle H}{\displaystyle |}}{C}{=}N\text{---}R_2$$

$$\underset{\text{sugar}}{} \qquad \underset{\text{protein}}{} \qquad \underset{\text{Schiff base}}{}$$

$$\Leftrightarrow R_1\text{---CO---CH}_2\text{---NH---}R_2 \rightarrow \rightarrow \rightarrow \text{AGEs}$$

$$\underset{\text{Amadori product}}{}$$

The Amadori product undergoes a series of higher-order reactions, the chemistry of which is not well understood, to produce yellow-brown fluorescent pigments that can cross-link proteins together (advanced glycation endproducts or AGEs). This overall process is also known as the Maillard or 'browning' reaction (see Monnier 1988). This can occur *in vivo* and is strongly implicated in several age-related changes, including cataracts, atherosclerosis, abnormalities in collagen and the thickening of basement membranes. Although these changes are more rapid in diabetics, it should be noted that glucose is one of the least reactive sugars. Deoxyribose is at least 200 times as reactive, ribose over 100 times, deoxyglucose about 25 times, and galactose about 5 times.

4.8 In Cattanach's translocation, a part of an autosomal chromosome is inserted into one arm of the X chromosome. It has two active genes that determine coat colour, one of which codes for tyrosinase. The same two genes on the pair of normal autosomes are mutant. The spreading of inactivation from the X chromosome to the translocated fragment results in an albino phenotype in female mice. However, these animals are mosaic, because in about half the cells the translocated chromosome remains active, so pigment is produced. Thus, there are patches of pigmented and unpigmented coat in a random mixture.

Another translocation can also be exploited, which is Searle's X autosome translocation. The X chromosome of this translocation is never inactivated, because that would result in an unbalanced genotype, which is lethal. These animals are therefore not mosaic, so if Cattanach's translocation is also present, this X chromosome is invariably inactivated, and the female animals are completely albino, as shown in Figure 4.6a. In both the mosaic and non-mosaic animals the pigment-forming genes became reactivated with age. Moreover there was evidence from the coat colour pattern that the two genes were reactivated sequentially rather than together (Cattanach 1974).

4.9 Migeon, Axelman & Beggs (1988) examined the frequency of HPRT$^+$ fibroblast clones from human females who are heterozygous HPRT$^+$/HPRT$^-$. Since HPRT is X linked, X chromosome reactivation during ageing would produce an excess of HPRT$^+$ clones, but only a slight and non-significant change in the ratio was seen. This, however, was a very insensitive assay because it could only detect several percent reactivation. In a much more sensitive assay, HPRT$^-$ clones were tested for reactivation to HPRT$^+$ after treatment with the demethylating agent deoxyazacytidine. It was found that reactivation was significantly more frequent in HPRT$^-$ clones from the older donors. This suggests that the methylation of the CpG island of HPRT gene on the inactive X chromosome may decline with age, which facilitates reactivation by deoxyazacytidine (for discussion, see Holliday 1989a).

4.10 'Prions' are infectious particles that do not contain any detectable nucleic acid. The gene for the major prion protein has been cloned, and presumably has some normal cellular function. It is thought that an altered form of this protein is able to propagate itself and produces by infection the so-called slow virus diseases, including kuru, scrapie and Creutzfeldt–Jakob syndrome. The symptoms are in the central nervous system, and the brain of affected individuals has some pathological features that are similar to those seen in Alzheimer's disease or senile dementias. It seems likely that normal and apparently uninfected brain tissue can, at least under some circumstances, be a source of infectious prion particles (for a review see Prusiner et al. 1993).

5.1 As cells approach Phase III, it seems that the population as a whole becomes senescent, as if the process was in some way synchronised. This is in part an illusion. As senescence approaches more and more cells cease to divide, so those that remain have to divide more than once to double the population. The effect of this is to compress the many cell divisions of the cycling cells into the last few population doublings (sometimes referred to as the 'concertina effect'). Instead of seeing patches of non-growing cells and colonies of growing cells, each trypsinisation of the culture evenly disperses the remaining growing cells, and the non-growing ones are passively transmitted. Eventually the culture becomes uniformly senescent and fails to achieve confluence.

5.2 Haematopoietic stem cells divide to give rise to further stem cells and to 'determined cells', that is, cells that will form all types of differentiated cells in blood. Determined cells divide prior to the final differentiation events. Thus there are at least two types of dividing cell in the population, and one of these – the determined cells – has limited proliferative capacity. In a transplant experiment a proportion of the total population is used, and only a minority of these are stem cells. With the use of a marker chromosome (a specific translocation), Ogden and Micklem (1976) followed a mixed population of normal and translocation-marked bone marrow cells. It was evident that after a few transplants the population was no longer mixed, but consisted entirely of one cell type. Subsequently the population died out. This result shows that the number of true stem cells that were being transplanted was very small, so by chance, the few remaining stem cells were of one type only. This strongly suggests that the finite lifespan of haematopoietic cells may simply be due to the dilution out of stem cells, which could be immortal (see also Note 2.2). The same dilution out of stem cells would not, of course, occur in the normal situation in bone marrow, where there are strict regulatory controls on their number and proliferation.

5.3 The commitment theory (Kirkwood & Holliday 1975b; Holliday et al. 1977, 1981) was originally proposed to explain the difference in formal terms between the finite growth of cultured diploid fibroblasts and the indefinite growth of transformed or partially transformed cell lines. Its assumptions are simple, and only three parameters are defined. It explains much existing data and makes quite specific predictions that have been verified experimentally. To understand the theory, it is necessary to outline its features in some detail, which is done here. It is not concerned with the actual mechanism of senescence, but with the structure of cell populations. It is important to be aware of these cell kinetics, because cells that are uncommitted or committed to eventual senescence occur *in vivo*, and it may be possible to obtain important information about these cells by experimental studies *in vitro*.

The theory accounts for the variability of the lifespan of different fibroblast populations, which has been well documented for strains such as W1-38 and MRC-5 (see Holliday et al. 1977). Imagine 100% uncommitted cells and a probability P of 25% that an uncommitted cell will give rise to a committed one at cell division. Then after one cell generation there are 75% uncommitted cells and 25% committed; after two divisions the proportions are 56.25% uncommitted and 43.75% committed; after three divisions the ratio becomes 42.2% to 57.8%, and so on. If all the cells are dividing at the same rate, the proportion of uncommitted cells declines exponentially. There is on average one uncommitted cell per million cells after about 47 generations. When the number of uncommitted cells becomes very small, the actual time the last one disappears becomes subject to random or stochastic variation. This has two important consequences. First, the last uncommitted cells will give rise to the final Phase III populations of senescent fibroblasts, since they produce the youngest committed cells, which have the greatest growth potential. Since, by chance, the last uncommitted cells will be lost by dilution at different times, the final lifespan of the population will also be variable. Computer simulation of fibroblast populations, in which $P = 0.275$, gives a good fit to actual observations (see Fig. 5 of Holliday et al. 1977).

The second consequence is that populations that exist as mixtures will often end up at Phase III with only one cell type. Zavala, Herner & Fialkow (1978) grew female fibroblast populations that had an approximately 1 : 1 mix of cells producing the isozymes G6PDA and G6PDB, which have different electrophoretic mobilities. The proportion of A and B isozymes can be measured by separating them on a gel and staining the respective enzyme activities. Although the title of their paper refers to selection of one or the other type as the cells approached senescence, the data actually show that the A and B cells often proliferate at the same rate for a major proportion of the lifespan. In many cases the ratio diverged from 1 : 1 only in Phase III, which is exactly what the commitment theory predicts. In populations of T lymphocytes, the cells produce many types of immunoglobulin on their surface. Yet in the final phase of growth, the population is clonal, since only one type of immunoglobulin is present (McCarron et al. 1987). These cells could all be derived from the longest surviving uncommitted cell. Similarly, in transplantation experiments haematopoietic cells *in vivo* show that mixed populations give rise to non-mixed ones (Ogden & Micklem 1976, and see Note 5.2).

According to the commitment theory a population has finite growth only if the committed cells are still dividing when the last uncommitted cells are lost. If committed cells become senescent and stop diving whilst uncommitted cells remain, then the population will keep growing indefinitely. This is illustrated in Figure to Note 5.3(a,b). When the first committed cells start dying, the growth rate of the whole population decreases, because less than 100% of cells are dividing. This change in growth rate has been consistently seen (Fig. to Note 5.3c) *provided* cells are split as soon as they become confluent, or as they approach confluence. (The growth rate clearly lessens if cells are incubated whilst confluent.) We referred to the period of rapid growth as Stage 1, the period of reduced growth rate as Stage 2, and the final senescent growth as Stage 3 (which corresponds to Hayflick's Phase III). The extent of reduction in growth rate provides a measure of P (see Kirkwood & Holliday 1975b). Also, Stage 2 should have non-cycling cells, whereas there should be none or few in Stage 1. This was confirmed experimentally (Fig. to Note 5.3c).

In real populations, the earliest passage or PD level could not contain a large proportion of uncommitted cells. For foetal lung strain MRC-5 at passage 8, it is estimated that only 0.005% uncommitted cells exist, that is, about 200 cells in a population of 2×10^6 cells. This means that for $P = 0.275$ the whole population of cells was uncommitted about 30 cell generations *earlier*. These 30 generations include the establishment of the initial primary culture (Phase I yields passage 1 cells), the subsequent 7 passages (in PDs) as well as the earlier growth of cells in the foetus. On the basis of simplifying assumptions, we calculate that there might be a population of uncommitted primordial fibroblasts (or precursors of fibroblasts) of about 10^{-3}–10^{-4} cells in the developing foetus (Holliday et al. 1977).

Most cell biologists would not expect that population size had anything to do with fibroblast lifespan, but in fact it does. The commitment theory predicts that a significant reduction in population size, or 'bottleneck', will have no effect on lifespan, provided the proportion of uncommitted cells is sufficiently high. However, as this proportion gets lower, a bottleneck will reduce lifespan by an amount that depends on the size of the bottleneck in relation to the size of the normal population. In bottleneck experiments with MRC-5, the observed reduction in lifespan was about eight PDs, which is the same as that predicted by

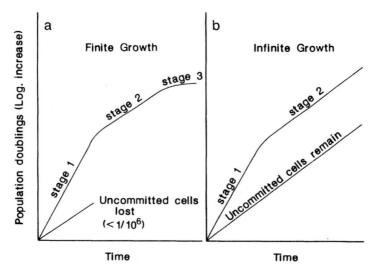

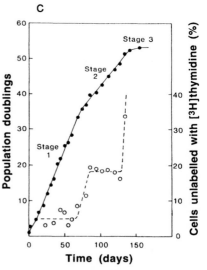

Figure for Note 5.3. (a,b) The cumulative growth of (a) normal diploid cells and (b) immortalised cells. The commitment theory proposes that the proportion of uncommitted cells continually declines, even though there is an absolute increase in cell numbers. If uncommitted cells are lost by dilution, then the remaining cells eventually die out. If, on the other hand, committed cells begin to die *before* the uncommitted cells are lost by dilution, then the rate of growth of the whole population is the same as the rate of increase of the uncommitted cells. This produces a steady state, and an immortal population. (c) The growth of human foetal lung fibroblast, strain MG4. In Stage 1 about 95% of the cells are cycling. The growth rate is slower at about PD35, because only about 80% of cells are cycling (Stage 2). Finally, growth slows further and ceases (Stage 3), and fewer and fewer cells become labelled (data not shown). In 29 parallel cultures, the change in growth rate between Stage 1 and Stage 2 was observed, but ^{3}H-thymidine labelling was only determined in one culture. (Part (c) reproduced with permission from Holliday, Huschtscha & Kirkwood, Further evidence for the commitment theory of cellular ageing, *Science* **213**, 1505–8. Copyright © AAAS.)

computer simulations. Initially, we assumed that bottlenecks should have no effect on lifespan once the last uncommitted cells had been lost. The experiments did not bear this out, as later bottlenecks reduced lifespan. But these experiments were done *before* the computer simulations! The latter demonstrated that these bottlenecks should indeed reduce lifespan by the amount observed, because the bottleneck eliminates the *youngest committed* cells, and these are the cells that have the greatest division potential. Bottlenecks fail to have an effect on lifespan when about 20 PDs of growth remain.

The important prediction that bottlenecks will have no effect on lifespan in very early cultures cannot be tested with MRC-5 cells, since they are not young enough. We therefore prepared primary Phase I cultures from foetal lung tissue, and subjected them to bottlenecks as soon as possible (Holliday et al. 1981). In these experiments with strain MG4, we observed a bimodal distribution of lifespans. Some had a significant reduction in lifespan in comparison to controls, and others did not. This also fits with the prediction if there is on average about one uncommitted cell per bottleneck population, which in this case is 10^4 cells. This cell will grow as the population is expanded to the normal size, but these cells will subsequently be diluted out by routine culture, and the lifespan will be in the normal range.

The final prediction of the theory is that very large populations should grow forever, because when the first committed cells die (beginning of Stage 2), there will still be uncommitted cells present, and these grow at the same rate as the whole population (Fig. to Note 5.3a,b). Unfortunately the experiment cannot be done unless facilities for growing 10^9–10^{10} cells become available. (Note that if such a population was divided up in separate containers, it would be necessary to bulk them all together every time the cells were passaged.) All the other predictions of the theory have stood up to experimental test. Perhaps that is surprising, since there are only three parameters in the model, P, M and N, and two of these are assumed to be constant ($P = 0.275$ and $M = 55$), which may be an unrealistic simplification.

It is known that individual clones of human fibroblasts have variable growth potential (Smith et al. 1978; Smith & Whitney 1980). The commitment theory predicts variable clonal lifespans, since individual cells can be anywhere along the pathway from initial commitment to senescence. However, it does not predict the bimodal distribution of clonal lifespans that has been documented (Smith & Whitney 1980) without at least one additional assumption, which has not yet been incorporated into the model.

The commitment model does not address the problem of the mechanism of cellular ageing, but it should be noted that it is compatible with the telomere hypothesis, if commitment is the loss of telomerase.

5.4 Molecular clocks that count cell divisions may be important in development, and could be responsible for certain quantitative traits (see Holliday 1991a). Cells are sometimes eliminated during development, so it is possible that a clock could trigger the cell suicide mechanism, known as apoptosis. Alternatively a developmental clock could be responsible for a regulatory switch in a developmental pathway. A clock based on the sequential methylation of DNA repeats is shown in the Figure to Note 5.4. Alternatively sequential loss of methylation could provide a counting mechanism, or there could be sequential base changes, from an A–T to a G–C base pair, or vice versa (Holliday & Pugh 1975). It is possible that there are examples of regulated programmed ageing based on any of these, or on similar mechanisms.

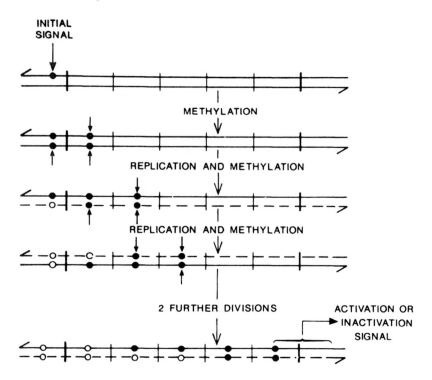

Figure for Note 5.4. A molecular model for a cell division counting mechanism or clock, essentially the same as that proposed by Holliday and Pugh (1975). A switch DNA methylase recognises and methylates one strand to the left of a series of repeated elements. This methylation signal is the substrate for a second enzyme, the 'clock methylase' that has the property of methylating both strands of the first of the repeated sequences but not further repeats (→●). It specifically requires a hemimethylated DNA substrate, which arises after DNA replication (- - -), and then the next repeat is methylated. Thus, there is a 'growing point' of methylation tied to cell division, and the sequences behind may or may not remain methylated (O). When all the repeats are methylated a signal is elicited, which may activate or repress an adjacent gene. (Reproduced with permission from Holliday 1991a.)

5.5 A problem arises in the replication of linear chromosomes: DNA cannot initiate DNA synthesis without a primer. The primer used consists of a short stretch of RNA complementary to the parental DNA template, which is later removed. Whereas the polymerase can completely replicate the end of one template, the other has a short unreplicated stretch (see Fig. to Note 5.5). On further replication a sequence is lost from the telomere, and with further cell divisions additional sequences are lost. The enzyme telomerase prevents this from happening, because it has the special ability to recognise the single-stranded overhang and fill in telomeric DNA. The enzyme contains RNA, which is complementary to the short repeated DNA telomere sequences.

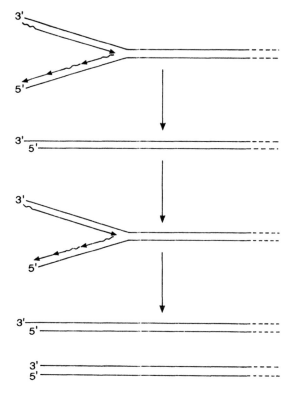

Figure for Note 5.5. The loss of telomeric DNA, which Olovnikov (1973) suggested could be a cause of cellular ageing. The wavy lines represent RNA primers, which are later removed. If the resulting gap is not filled, then after the next replication one of the daughter chromosomes has lost terminal DNA. When the enzyme telomerase is present the gap, or overhang, is filled with new DNA.

5.6 Mutations in somatic cells are usually detected by plating cells in a selective medium and looking for rare mutant colonies that can survive in the medium. Thus cells lacking the enzyme hypoxanthine-guanine phosphoribosyl transferase (HPRT) can grow in the presence of the base analogue 6-thioguanine. Obviously it is not possible to grow colonies of senescent cells, so the method cannot be applied. At best, one can look at the first two-thirds or so of the lifespan, as Gupta (1980) did. He found no significant increase in mutation during this period. Another mutation assay depends on the screening of individual cells for a specific phenotype. Some mutant forms of G6PD can use deoxyglucose in a substrate, whereas the wild-type enzyme has only weak activity. A histochemical assay can reveal which cells have a mutant enzyme. When this method was used, heavily stained cells were seen amongst a background of non-staining cells (Fulder & Holliday 1975). However, these cells simply have an elevated level of G6PD, rather than an altered substrate specificity. The gene for G6PD is subject to regulatory control, so it is possible that the variants seen are due to mutations in regulatory genes, although no direct evidence for this was obtained. In any event, it was clearly shown that the frequency of cell variants increased rapidly in Phase III, with exponential kinetics.

5.7 Pereira-Smith & Smith (1982) reported that senescent human cells fused to young cells produced a phenotype that is senescent. However, in earlier experiments Hoehn, Bryant & Martin (1978) examined the replicative lifespan of tetraploid hybrids between parents at different PD levels, and found that it was in general intermediate between those of the parents. Senescence in fungi is determined by cytoplasmic events and is dominant in heterokaryons. This led to the proposal that the senescence might be due to an error catastrophe in protein synthesis (Holliday 1969), which would be dominant in a mixed cytoplasm. Some indirect evidence for this was obtained (see Note 6.5). However, subsequent studies of senescence in *Podospora* show that mtDNA is abnormal, and such mitochondria are suppressive or dominant in heterokaryons (see Note 6.4). One might expect that there would always be selection for cells with normal mitochondria, as presumably happens in all fungi that do not show senescence. The reasons for the appearance and rapid spread of abnormal mitochondria in *Podospora* are not apparent.

5.8 At a workshop on the *Control of Cell Proliferation in Senescent Cells,* organised by H. Warner and E. Wang, Montreal, 12–13 August 1988 (see Warner & Wang 1989), a substantial number of speakers discussed the regulation of senescence, especially the DNA inhibitor theory. This stimulated an assessment and critique of the view that senescence is regulated: 'The limited proliferation of cultured human diploid cells: Regulation or senescence?' (Holliday 1990) and an accompanying defence of the inhibitor hypothesis (Smith 1990).

5.9 Evidence has been obtained that human diploid cells infected with SV40 continue to lose telomeric DNA until crisis is reached (Counter et al. 1992). The emergence of rare immortalised clones is thought to be due to the acquisition of telomerase activity. This is an alternative explanation of immortalisation: it would be a regulatory change, but not related to the bypass of an inhibitor of DNA synthesis.

It is important to know whether the crisis of SV40-infected cells is the same or different from the senescence of normal diploid cells. The telomere hypothesis suggests that the reason for the cessation of growth is the same in both cases. However, there are important differences between crisis and senescence. In the former, cells continue to divide, but they detach and float in the medium. Also, DNA methylation is maintained in pre-crisis cells (Matsumura et al. 1989), whereas it gradually declines in diploid cells.

5.10 A deleterious recessive mutation in an autosomal gene will, by definition, have no phenotypic effect. It might, however, become homozygous by mitotic crossing-over or gene conversion. Alternatively, the single remaining wild-type gene might also undergo mutation, although the probability of two homologous mutations in a single cell will be very low. The X chromosome is present in single copy in males, and is effectively single in females as well, owing to X chromosome inactivation. On the basis of such assumptions, Holliday & Kirkwood (1981) calculated the probability of populations of human fibroblasts dying out as a result of the accumulation of recessive mutations that inactivated single genes. To produce the 'Hayflick limit' to growth, a very high mutation rate (10^{-3}–10^{-4}/ gene/cell generation) would be required, and also one would expect a much higher frequency of non-viable cells in Phase II than is actually observed. It is interesting to note that Szilard (1959), in his elaboration of the somatic mutation theory of ageing, concluded genes were too small a target, and that it was only possible to develop a viable theory if whole chromosomes were inactivated or lost.

Codominant or dominant mutations, by definition, have an immediate effect on the phenotype, so their existence has been invoked by more recent proponents of the somatic mutation theory (e.g. Morley 1982). Since a large proportion of all known inherited human defects are dominant (McKusick 1990), it is not unreasonable to suppose that the same could be true for somatic cells. In the case of dividing cells, mutations that prevent or slow down growth would be selected out, but they could simply accumulate in non-dividing cells.

5.11 Autofluorescent material, perhaps the product of lipid peroxidation, steadily accumulates during the cumulative growth of human diploid fibroblasts (Rattan et al. 1982), and there is more autofluorescence in Werner's syndrome cells, which have limited growth potential. These experiments utilised a fluorescence-activated cell sorter (FACS), and the same instrument can be used to select out the least fluorescent cells. When this was done with normal MRC-5 cells, the level of fluorescence remained constant, but the cells had a normal lifespan in PDs (L. I. Huschtscha, unpublished observations).

6.1 It was first suggested by Orgel (1963) that suicide proteins might exist: 'One can also conceive of special mechanisms of quality control; for example, special proteins might be synthesised which are converted by a certain class of errors into lethal polypeptides.' This idea was taken further with a specific suggestion that the introduction, by error, of a cysteine residue in a particular protein domain might drastically change the structure of a protein by forming an internal –S–S– bridge with an –SH group already present (Holliday 1975). A suicide protein could kill cells by binding strongly to the promoter region of an indispensable gene. If there are a large number of 'quality control' molecules, then any cell that was error prone could make a small number of suicide protein molecules.

6.2 In genomic imprinting, maternal and paternal chromosomes transmit different information, which in some way is superimposed on the normal genetic information. This imprint is subsequently erased, and chromosomes are again imprinted in the next generation. Imprinting can be regarded as a form of complementation, since at particular loci it is necessary to have both paternal and maternal genes to achieve normal development. Two paternal or two maternal genes produce developmental abnormalities. It is likely that some imprinted genes are inactive in paternal chromosomes, and others are inactive in the maternal chromosomes; thus, the two haploid genomes complement each other.

There is growing evidence that imprinting is due to DNA methylation, and that it provides an important example of epigenetic inheritance of a given pattern of methylation. The imprint would be erased by the later removal of methylation, followed by the imposition of the new pattern required for the production of normal offspring (for reviews, see Monk & Surani 1990; Jost & Saluz 1993).

6.3 The size of organisms is constrained by the relationship between length and weight. For a cube, the mass or weight increases as the third power of the length. A comparable relationship exists for all solid three-dimensional objects, and this sets limits on the evolution of increasing size, or the increasing growth of organisms such as large trees.

6.4 The senescence of *Podospora* is associated with abnormalities in mitochondrial DNA. In particular, SEN-DNA frequently consists of an amplified portion (α) of the mitochondrial genome, and senescence is strictly correlated with the presence of SEN-DNA molecules. For references to the many studies that have been carried out in three laboratories, see Belcour & Vierny (1986), who write: 'The causal relationship between senescence and modifications of mtDNA is not definitely established and the molecular mechanisms by which SEN-DNA molecules are generated from the mitochondrial chromosome remain obscure.'

6.5 The same methods used with the *leu*-5 strain of *Neurospora* were also applied to *nd*. As the culture aged, temperature-sensitive glutamic dehydrogenase was produced, and also inactive cross-reacting material (Lewis & Holliday 1970). In earlier studies, *nd* was combined in haploid strains with an adenine auxotroph (*ad*-3), which was known to be phenotypically reverted by RNA base analogues, such as fluorouracil or azaguanine. In other words, the introduction of errors in RNA synthesis allowed the *ad*-3 strain to grow on unsupplemented medium. It was found that *nd* also suppressed *ad*-3, suggesting that the *nd* strain has a high error level in protein synthesis (Holliday 1969). In other experiments it was shown that amino acid analogues significantly reduced the natural lifespan of *Podospora*, whereas a low level of the protein synthesis inhibitor cycloheximide had no effect. The relationship between these observations and abnormal mtDNA in senescent *Podospora* (see Note 6.4) is far from clear.

7.1 DNA synthesis can be measured by the incorporation of ^{3}H-thymidine into cell nuclei, and the preparation of autoradiographs. If the label is present when chromosomes are being replicated, the nuclei are heavily labelled, but if the label is present outside the normal period of DNA replication, no nuclear grains are seen in autoradiographs. This situation changes if the DNA is damaged, for example by UV light, because then the damage is excised, and in filling the gap some thymidine is incorporated. This low-level labelling is called 'unscheduled DNA synthesis' (UDS), and it can provide a quantitative measure of the amount of repair.

7.2 Many studies on the relationships between longevity and various biological parameters have relied on early 'data bases', that is, the recorded longevity of captive animals, mainly in zoos. Some of these early records are not very reliable, because the age of the animal acquired by the zoo may not be known. Also, veterinary care was much less efficient than it is in modern zoos, so the death of animals may have been from a particular disease rather than natural ageing. In general, the recorded longevities of captive animals have increased throughout the century, but in almost all cases the longevities of only small numbers of each species are available. The best records have been compiled by Marvin L. Jones, the Registrar of San Diego Zoo, California.

7.3 Gavrilov & Gavrilova (1991) list 53 references in this century to maximum human lifespan, ranging from 60 years to 167 years. Interestingly, one paper by a well-known gerontologist uses no less than 10 figures, from 86 to 115 years. The maximum documented lifespan is 120 years and 237 days, in the *Guinness Book of Records*.

7.4 For a time span of fifty thousand years at 4.5 generations per century there are 2250 generations. If the increase in the number of individuals is 500-fold, then the average increment in population size x per generation is given by the formula $x^{2250} \approx 500$, then $x \approx 1.0028$ or 0.28% per generation.

This value will be slightly lower if the founder population was greater, or if the number of generations was greater. On the basis of any reasonable assumptions, the average rate of increase in the population must have been extremely low, but there would have been major fluctuations in this rate, depending on food supply and other environmental conditions.

7.5 A study in East Africa indicates that the average interbirth interval for lactating women is 27 months, and 15 months for non-lactating women (Saxton & Servadda 1969). Thus, the number of offspring is in part determined by the survival or death of infants. In chimpanzees and gorillas in their natural environments, infant mortality is about 25% (Harcourt, Fossey & Sabater-Pi 1981; Courtenay & Santow 1989). If we assume the same value for human infants in a hunter-gatherer community, then females will produce on average one offspring every two years. One also has to take into account the survival of female children to breeding age and the mortality of breeding females, including death in childbirth. This introduces several imponderables, but if one assumes that the population renews itself without increase, then it is possible to estimate (1) the average number of offspring per female after they reach reproductive age, and (2) the expectation of life. The former value is about six offspring, and the expectation of life of females who reach 16 years of age is 28 years; the expectation of life at birth is only 15 years. (It should be noted that if infanticide is practised as a means of population control, the average interbirth interval would be reduced. The main effect would be to increase the number of offspring born.)

These estimates are based on an annual mortality rate of 7% after the age of 1 year. This is an oversimplification, as it assumes an exponential survival curve, which is probably unrealistic (see Chapter 1). However, from the analysis of skeletons in graveyards in ancient Greece (Angel 1947), it has been estimated that survival was not far removed from an exponential loss of life, whereas one might have expected otherwise in a well-organised society.

7.6 According to the assumptions made in Note 7.5, namely 25% infant mortality and a 7% annual mortality rate, then about 3% of the female population would reach the age of 45. However, these women would be those who had produced the most offspring in the community. The kin-selective force for non-reproduction would not be trivial, given the possible number of children and grandchildren who might benefit from this trait.

8.1 It is very often assumed that DNA-damaging agents are carcinogens because they produce mutations. What is not so widely realised is that DNA damage could also damage the normal pattern of DNA methylation, and in some cases this could induce aberrant changes in gene expression. One way in which the repair of DNA damage could result in the loss of DNA methylation is shown in Figure for Note 8.1. It is also possible that misrepaired DNA damage near a methylated CpG could inhibit normal DNA methylase activity. Thus, epigenetic changes following DNA damage could be important in tumour progression (see also Holliday 1987).

Some changes leading to malignancy are reversible, suggesting that they have an epigenetic basis. For example, embryonal tumour cells, or teratocarcinomas, can be produced by placing a mouse embryo under the kidney capsule of an adult animal. These cells are malignant, but if individual cells are injected into mouse embryos, a chimaera can be formed that contains a variety of differentiated tissues derived from the teratocarcinoma. The cells may not be completely normal, but the malignant phenotype has certainly been reversed.

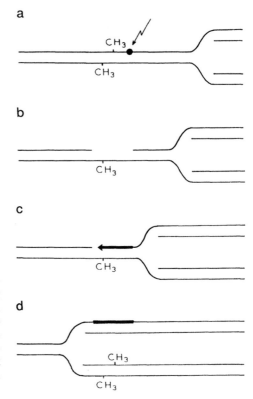

Figure for Note 8.1. Damage to DNA in front of an advancing replication fork can result in loss of DNA methylation, if the patch filled by repair synthesis is not methylated. Similarly, damage behind the replication fork can lead to loss of methylation (see Holliday 1979).

9.1 The pyrimidine analogue 5-fluorouracil (FU) is incorporated into RNA, and it was shown many years ago that it could induce translational errors. Certain mutants of bacteriophage T4 can be phenotypically suppressed by FU, because the errors are sufficiently high to produce some wild-type protein, even though the DNA is mutant (see also Note 6.5). The amino glycoside antibiotic paromomycin (Pm) binds to eukaryotic ribosomes and reduces the accuracy of translation of mRNA. The most detailed studies have been done in yeast, where it has been shown that certain mutants are phenotypically suppressed by Pm.

5-azacytidine and azadeoxycytidine (5AC) are potent inhibitors of DNA methylation. They are incorporated into DNA, and it is probable that DNA methyltransferase binds covalently to the cytosine analogue, thus preventing the enzyme from carrying out its normal function. A single 24-hour dose of 5AC early in the growth of human fibroblasts has a dramatic effect in decreasing lifespan. The cells 'remember' that they have been treated, as the effect is seen many weeks and passages later. This suggested that the decline in DNA methylation, which is accelerated by a single 5AC treatment, may be an important cause of ageing (see references in Table 9.1). Subsequent studies with SV40-infected human cells made this possibility less likely. These cells do not lose methylation, but they do have a finite lifespan that is terminated by a 'crisis' (Matsumura et al. 1989) (see also Note 5.9).

Human fibroblasts can grow at 40°, but their lifespan is greatly reduced. This is probably due to premature ageing, because transfer of cells back to 37° does not lead to a full recovery of growth potential. Cells can grow slowly at 32° and their lifespan is also short. However, in this case transfer of cells from 32° to 37° results in total recovery, suggesting that the low temperature induces an abnormal physiological state that is reversible. This study also showed that cells grown at 34° have the same lifespan as those grown at 37°. It has sometimes been suggested that reduction of the temperature of homothermic mammals would increase longevity, but this is not true for cell populations.

9.2 The Amadori reaction between the ε-amino group of lysine and sugar aldehydes is described in Note 4.7. It was discovered that the β-amino group of the dipeptide carnosine reacts with sugars more rapidly than does lysine (Michaelis & Hipkiss 1994). This suggested that the role of carnosine in cells may be to react with sugars to prevent the non-enzymic glycosylation of proteins. One important sugar that reacts with lysine and more rapidly with carnosine is deoxyribose. Spontaneous apurinic gaps are common in DNA, and the exposed deoxyribose in its aldehyde form is a potential site for DNA–protein cross-linking (mediated by a lysine side chain). Carnosine could have a role in repair if it prevents DNA–protein cross-linking, by reacting preferentially with apurinic sites.

9.3 The *recA* mutation in *E. coli* is a classical example of a pleiotropic mutant phenotype. When DNA is damaged in normal cells, the amount of *recA* protein increases. This is required for recombinational repair, but *recA* protein also cleaves the *lex* repressor protein. *Lex* controls a *regulon:* a collection of about 20 different genes all of which are repressed by *lex* protein and induced when it is inactivated. This constitutes the cells' SOS response to DNA-damaging agents. In the absence of *recA* function, not only is there no recombination, but none of the SOS response genes are induced. Thus, the activity of one gene affects the expression of 20 or so other genes (see Friedberg 1985; Sedgwick 1986).

References

Adams, R. L. P. (1990). DNA methylation. The effect of minor bases on DNA–protein interactions. *Biochem. J.* **265**, 309–20.

Adelman, R., Saul, R. L., & Ames, B. N. (1988). Oxidative damage to DNA: Relation to species metabolic rate and lifespan. *Proc. Natl. Acad. Sci. USA* **85**, 2706–8.

Aizawa, S., & Mitsui, Y. (1979). A new cell surface marker of aging in human diploid fibroblasts. *J. Cell. Physiol.* **100**, 383–8.

Allsopp, R. C., Vaziri, H., Patterson, C., Goldstein, S., Younglai, E. V., Futcher, A. B., Greider, C. W., & Harley, C. B. (1992). Telomere length predicts replicative capacity of human fibroblasts. *Proc. Natl. Acad. Sci. USA* **89**, 10114–18.

Alterman, M., Carvan, M., Srivastava, V., Leakey, J., Hart, R., & Busbee, D. (1993). Effects of aging and long term caloric restriction on hepatic micro-somal homooxegenases in female Fischer 344 rats: Alterations in basal cytochrome P450 catalytic activities. *Age* **16**, 1–8.

Ames, B. N. (1983). Dietary carcinogens and anticarcinogens: Oxygen radicals and degenerative diseases. *Science* **221**, 1256–64.

Ames, B. N., & Gold, L. S. (1991). Endogenous mutagens and the causes of aging and cancer. *Mutat. Res.* **250**, 3–16.

Ames, B. N., Saul, R. L., Schwiers, E., Adelman, R., & Cathcart, R. (1985). Oxidative damage as related to cancer and aging: The assay of thymine glycol, thymidine glycol and hydroxymethyl uracil in human and rat urine. In *Molecular Biology of Aging*, ed. R. S. Sohal, L. S. Birnbaum & R. G. Cutler, pp. 137–44. Raven Press, New York.

Ames, B. N., & Shigenaga, M. K. (1992). Oxidants are a major contribution to aging. *Ann. New York Acad. Sci.* **663**, 85–96.

Ames, B. N., Shigenaga, M. K., & Hagen, T. M. (1993). Oxidants, antioxidants and the degenerative diseases of aging. *Proc. Natl. Acad. Sci. USA* **90**, 7915–22.

Anasimov, V. N. (1986). Relevance of age to some aspects of carcinogenesis. *IARC Sci. Publ.* **58**, 115–26.

(1991). Effects of factors prolonging lifespan on carcinogenesis. *Ann. New York Acad. Sci.* **621**, 373–84.

Angel, J. L. (1947). The length of life in ancient Greece. *J. Gerontol.* II, **1**, 18–24.

Austad, S. N. (1993). Retarded senescence in insular population of Virginia Opossums (*Didelphis virginiana*). *J. Zool.* **229**, 695–708.

Balasubramanian, D., Bhat, K. S., & Rao, G. N. (1990). Factors in the preva-

173

lence of cataract in India: Analysis of the recent Indo–U.S. study of age-related cataracts. *Curr. Sci.* **59**, 498–505.

Beckman, K. B., Hagen, T. M., & Ames, B. N. (1992). DNA oxidation and the extranuclear mutation theory of aging. (Abstr.) In *The Molecular Biology of Aging*, p. 1, Cold Spring Harbor Laboratory, New York.

Belcour, L., & Vierny, C. (1986). Variable DNA splicing sites of a mitochondrial intron: Relationship to the senescence process in *Podospora*. *EMBO J.* **5**, 609–14.

Bell, G. (1988). *Sex and Death in Protozoa*. Cambridge University Press, Cambridge.

Ben-Ishai, R., & Peleg, L. (1975). Excision in primary cultures of mouse embryo cells and its decline in progressive passages and established cell lines. In *Molecular Mechanisms for Repair of DNA*, ed. P. C. Hanawalt and R. B. Setlow, pp. 607–10. Plenum, New York.

Benditt, E. P., & Benditt, J. M. (1973). Evidence for a monoclonal origin of human atherosclerotic plaques. *Proc. Natl. Acad. Sci. USA* **70**, 1753–6.

Bernstein, C. (1981). Why are babies born young? Meiosis may prevent aging of the germ line. *Perspect. Biol. Med.* **22**, 539–44.

Bernstein, C., & Bernstein, H. (1991). *Aging, Sex and DNA Repair*. Academic Press, San Diego.

Bierman, E. I. (1978). The effect of donor age on the *in vitro* life span of cultured human arterial smooth-muscle cells. *In Vitro* **14**, 951–5.

Bird, A. P. (1986). CpG-rich islands and the function of DNA methylation. *Nature* **321**, 209–13.

Bradley, M. O., Hayflick, L., & Schimke, R. T. (1976). Protein degradation in human fibroblasts (W1–38): Effects of ageing, viral transformation and amino acid analogues. *J. Biol. Chem.* **251**, 3521–9.

Branscomb, E. W., & Galas, D. J. (1975). Progressive decrease in protein synthesis accuracy induced by streptomycin in *Escherichia coli. Nature* **254**, 161–3.

Breitbart, R. E., Andreadis, A., & Nadal-Ginard, B. (1987). Alternative splicing: A ubiquitous mechanism for the generation of multiple protein isoforms from single genes. *Annu. Rev. Biochem.* **56**, 467–95.

Brown, S., & Rastan, S. (1988). Age related reactivation of an X linked gene close to the inactivation centre in the mouse. *Genet. Res.* **52**, 151–4.

Bruce, S. A., & Deamond, S. F. (1991). Longitudinal study of *in vivo* wound repair and *in vitro* cellular senescence of dermal fibroblasts. *Exp. Gerontol.* **26**, 17–27.

Bruce, S. A., Scott, F. D., & Ts'o, P. O. P. (1986). *In vitro* senescence of Syrian hamster mesenchymal cells of fetal to aged adult origin: Inverse relationship between *in vivo* donor age and *in vitro* proliferative capacity. *Mech. Ageing Dev.* **34**, 151–73.

Brunk, U., Ericsson, J. L. E., Ponten, J., & Westermark, B. (1973). Residual bodies and 'aging' in cultured human glia cells. *Exp. Cell Res.* **79**, 1–14.

Bunn, C. L., & Tarrant, G. M. (1980). Limited lifespan in somatic cell hybrids and cybrids. *Exp. Cell Res.* **127**, 385–96.

Bush, A. I., Beyreuther, K., & Masters, C. L. (1992). βA4 amyloid protein and its precursor in Alzheimer's disease. *Pharmacol. Ther.* **56**, 97–117.

Burch, P. R. J. (1968). *An Inquiry Concerning Growth, Disease and Ageing*. Oliver & Boyd, Edinburgh.

Burnet, F. M. (1974). *Intrinsic Mutagenesis: A Genetic Approach to Aging*. Wiley, New York.

Cairns, J. (1978). *Cancer: Science and Society.* W. H. Freeman, San Francisco.

Carey, J. R., Liedo, P., Orozco, D., & Vaupel, J. W. (1992). Slowing of mortality rates at older ages in large medfly cohorts. *Science* **258,** 457–61.

Cattanach, B. L. (1974). Position effect variegation in the mouse. *Genet. Res.* **23,** 291–306.

Cerami, A. (1986). Ageing of proteins and nucleic acids: What is the role of glucose? *Trends Biochem. Sci.* **11,** 311–14.

Chandrasekhar, S., Norton, E., Millis, A. J. T., & Izzard, C. S. (1983). Functional changes in cellular fibronectin from late passage fibroblasts *in vitro. Cell Biol. Int. Rep.* **7,** 11–22.

Charlesworth, B. (1980). *Evolution in Age-Structured Populations.* Cambridge University Press, Cambridge.

Child, C. M. (1915). *Senescence and Rejuvenescence.* Chicago University Press, Chicago.

Comfort, A. (1979). *The Biology of Senescence,* 3rd Ed. Churchill Livingstone, London.

Counter, C. M., Avilion, A. A., LeFerve, C. E., Stewart, N. G., Greider, C. W., Harvey, C. B., & Bacchetti, S. (1992). Telomere shortening associated with chromosome instability is arrested in immortal cells which express telomerase activity. *EMBO J.* **11,** 1921–9.

Court Brown, W. M., Buckton, K. E., Jacobs, P. A., Tongh, I. M., Kuenssberg, E. V., & Knox, J. D. E. (1966). *Chromosome Studies on Adults, Eugenics Laboratory Memoir XLII.* Cambridge University Press, Cambridge.

Courtenay, J., & Santow, G. (1989). Mortality of wild and captive chimpanzees. *Folia Primatol.* **52,** 167–77.

Cristofalo, V. J. (1974). Metabolic aspects of aging in diploid human cells. In *Aging in Cell and Tissue Culture,* ed. E. Holeckova & V. J. Cristofalo, pp. 83–119. Plenum, New York.

Cristofalo, V. J., & Kabakjian, J. (1975). Lysosomal enzymes and aging *in vitro:* Subcellular enzyme distribution and effect of hydrocortisone on cell lifespan. *Mech. Ageing Dev.* **4,** 19–28.

Cristofalo, V. J., Parris, N., & Kritchevsky, D. (1967). Enzyme activity during the growth and aging of human cells *in vitro. J. Cell Physiol.* **69,** 263–71.

Cristofalo, V. J., & Sharf, B. B. (1973). Cellular senescence and DNA synthesis. *Exp. Cell Res.* **76,** 419–27.

Croute, F., Vidal, S., Soleilhavoup, J. P., Vincent, C., Serre, G., & Planel, H. (1986). Effects of very low dose rate of chronic ionising radiation on the division of potential of human embryonic lung fibroblasts *in vitro. Exp. Gerontol.* **21,** 1–11.

Curtis, H. J. (1966). *Biological Mechanisms of Ageing.* Thomas, Springfield, Ill.

Curtsinger, J. W., Fukui, H. H., Townsend, D. R., & Vaupel, J. W. (1992). Demography of genotypes: Failure of the limited lifespan paradigm in *Drosophila melanogaster. Science* **258,** 461–3.

Cutler, R. G. (1975). Evolution of human longevity and the genetic complexity governing aging rate. *Proc. Natl. Acad. Sci. USA* **72,** 4664–8.

(1982). The dysdifferentiation hypothesis of mammalian aging and longevity. In *The Aging Brain: Cellular and Molecular Mechanisms of Aging in the Nervous System,* ed. E. Giacobini, F. Filogamo, G. Giacobini & V. A. Vernadakis, pp. 1–19. Raven, New York.

(1984). Carotenoids and retinol: Their possible importance in determining longevity of primate species. *Proc. Natl. Acad. Sci. USA* **81,** 7627–31.

(1985). Peroxide producing potential of tissues: Inverse correlation with longevity of mammalian species. *Proc. Natl. Acad. Sci. USA* **82**, 4798–802.

Daniel, C. W., DeOme, K. B., Young, J. T., Blair, P. B., & Faulkin, L. J. (1968). The *in vivo* lifespan of normal and preneoplastic mouse mammary glands: A serial transplantation study. *Proc. Natl. Acad. Sci. USA* **61**, 53–60.

Daniel, C. W., & Young, L. J. T. (1971). Influence of cell division on an aging process: Lifespan of mouse mammary epithelium during serial propagation *in vivo*. *Exp. Cell Res.* **65**, 27–32.

Davies, K. J. A., & Goldberg, A. L. (1987). Proteins damaged by oxygen radicals are rapidly degraded in extracts of red blood cells. *J. Biol. Chem.* **262**, 8227–34.

Davies, T. F., Platzer, M., Schwartz, A. E., & Friedman, E. W. (1985). Short- and long-term evaluation of normal and abnormal human thyroid cells in monolayer culture. *Clin. Endocrinol.* **23**, 469–79.

de Beer, G. (1958). *Embryos and Ancestors*, 3rd. Ed. Oxford University Press, Oxford.

Dean, R. T., Gebicki, J., Gieseg, S., Grant, A. J., & Simpson, J. A. (1992). Hypothesis: A damaging role in aging for reactive protein oxidation products. *Mutat. Res.* **275**, 387–93.

Dean, R. T., Gieseg, S., & Davies, M. J. (1993). Reactive species and their accumulation on radical-damaged proteins. *Trends Biochem. Sci.* **18**, 437–41.

Dempsey, J. L., Pfeiffer, M., & Morley, A. A. (1993). Effect of dietary restriction on *in vivo* somatic mutation in mice. *Mutat. Res.* **291**, 141–5.

Diamond, J. M. (1982). Big bang reproduction and ageing in male marsupial mice. *Nature* **298**, 115–16.

Dice, J. F. (1987). Molecular determinants of protein half lives in eukaryotic cells. *FASEB J.* **1**, 349–57.

Dingley, F., & Maynard Smith, J. (1969). Absence of a life-shortening effect of amino acid analogues on adult *Drosophila*. *Exp. Gerontol.* **4**, 145–9.

DiPaulo, J. A., Popescu, N. C., Alvarez, L., & Woodworth, C. D. (1993). Cellular and molecular alterations in human epithelial cells transformed by recombinant human papillomavirus DNA. *Crit. Rev. Oncogenesis* **4**, 337–60.

Doerfler, W. (1993). Pattern of *de novo* methylation and promoter inhibition: Studies on the adenovirus and the human genome. In *DNA Methylation: Molecular Biology and Biological Significance*, ed. J. P. Jost & H. P. Saluz, pp. 262–99. Birkhauser, Basel.

Drake, J. M. (1970). *The Molecular Basis of Mutation*. Holden Day, San Francisco.

Dykhuizen, D. (1974). The evolution of cell senescence, atherosclerosis and benign tumours. *Nature* **251**, 616–18.

Edelmann, P., & Gallant, J. (1977a). Mistranslation in *E. coli*. *Cell* **10**, 131–7.
    (1977b). On the translational error theory of aging. *Proc. Natl. Acad. Sci. USA* **74**, 3396–8.

Edick, G. F., & Millis, A. J. T. (1984). Fibronectin distribution on the surfaces of young and old human fibroblasts. *Mech. Ageing Dev.* **20**, 243–52.

Egilmez, N. K., Chen, J. B., & Jazwinski, S. M. (1989). Specific alteration in transcript prevalence during the yeast lifespan. *J. Biol. Chem.* **264**, 14312–17.

Ellis, R. J., & van der Vies, S. M. (1991). Molecular chaperones. *Annu. Rev. Biochem.* **60**, 321–347.

Epstein, C. J., Martin, G. M., Schultz, A. L., & Motulsky, A. G. (1966). Werner's syndrome: A review of its symptomatology, natural history, pathological features, genetics and relationship to the natural ageing process. *Medicine (Baltimore)* **45**, 177–221.

Everitt, A. [V.], & Meites, J. (1989). Aging and anti-aging effects of hormones. *J. Gerontol. Biol. Sci.* **44**, 139–47.

Everitt, A. V., Olsen, G. G., & Burrows, G. R. (1968). The effect of hypophysectomy on the aging of collagen fibers in the tail tendon of the rat. *J. Gerontol.* **23**, 333–6.

Everitt, A. V., Seedsman, N. J., & Jones, F. (1980). The effects of hypophysectomy and continuous food restriction on collagen ageing, proteinuria, incidence of pathology and longevity in the male rat. *Mech. Ageing Dev.* **12**, 161–72.

Fairweather, S., Fox, M., & Margison, G. P. (1987). The *in vitro* lifespan of MRC-5 cells is shortened by 5-azacytidine induced demethylation. *Exp. Cell Res.* **168**, 153–9.

Finch, C. E. (1987). Neural and endocrine determinants of senescence: Investigation of causality and reversibility by laboratory and clinical interventions. In *Modern Biological Theories of Aging*, ed. H. R. Warner, R. N. Butler, R. L. Spratt & E. L. Schneider, pp. 261–306. Raven Press, New York.
(1990). *Longevity, Senescence and the Genome*. University Press, Chicago.

Fraga, C. G., Shigenawa, M. K., Park, J.-W., Degan, P., & Ames, B. N. (1990). Oxidative damage to DNA during aging: 8 hydroxy–2´-deoxyguanosine in rat organ DNA and urine. *Proc. Natl. Acad. Sci. USA* **87**, 4533–7.

Francis, A. A., Lee, W. H., & Reagan, J. D. (1981). The relationship of DNA excision repair of ultraviolet-induced lesions to the maximum lifespan of mammals. *Mech. Ageing Dev.* **16**, 181–9.

Friedberg, E. C. (1985). *DNA Repair*. W. H. Freeman, San Francisco.

Friedman, D. B., & Johnson, T. E. (1988). A mutation in the *age* 1 gene of *Caenorhabditis elegans* lengthens life and reduces hermaphrodite fertility. *Genetics* **118**, 75–86.

Fukuchi, K., Tanaka, K., Kumahara, Y., Marumo, K., Pride, M. B., Martin, G. M., & Monnat, R. J. (1990). Increased frequency of 6-thioguanine resistant peripheral blood lymphocytes in Werner's syndrome patients. *Hum. Genet.* **84**, 249–52.

Fukuchi, K., Tanaka, K., Nakura, J., Kumuhara, Y., Uchida, T., & Okada, Y. (1985). Elevated spontaneous mutation rate in SV40 transformed Werner syndrome fibroblast cell lines. *Somatic Cell Mol. Genet.* **11**, 303–8.

Fulder, S. J. (1978). Spontaneous mutations and ageing of human cells in culture. *Mech. Ageing Dev.* **10**, 101–15.

Fulder, S. J., & Holliday, R. (1975). A rapid rise in cell variants during the senescence of populations of human fibroblasts. *Cell* **6**, 67–73.

Fulder, S. J., & Tarrant, G. M. (1975). Possible changes in gene activity during the aging of human fibroblasts. *J. Exp. Gerontol.* **10**, 205–11.

Ganguly, T., & Duker, N. J. (1992). Reduced 5 hydroxymethyluracil DNA glycosylase activity in Werner's syndrome cells. *Mutat. Res.* **275**, 87–96.

Gardiner-Garden, M., & Frommer, M. (1987). CpG islands in vertebrate genomes. *J. Mol. Biol.* **196**, 261–82.

Gavrilov, L. A., & Gavrilova, N. S. (1991). *The Biology of Life Span: A Quantitative Approach*. Harwood, London.

Gartler, S. M., & Riggs, A. D. (1983). Mammalian X chromosome inactivation. *Annu. Rev. Genet.* **17**, 155–90.

Gething, M. J., & Sambrook, J. (1992). Protein folding in the cell. *Nature* **355**, 33–45.

Goodman, S. A., & Makinodan, T. (1975). Effect of age on cell-mediated immunity in long-lived mice. *Clin. Exp. Immunol.* **19**, 533–542.

Goto, S., Ishigami, A., & Takahashi, R. (1990). Effect of age and dietary restriction on accumulation of altered proteins and degradation of proteins in mouse. In *Liver and Aging*, ed. K. Kitain, pp. 137–47. Elsevier, Amsterdam.

Grimley Evans, J., & Franklin Williams, T. (eds.) (1992). *Oxford Textbook of Geriatric Medicine*. Oxford University Press, Oxford.

Grist, S. A., McCarron, M., Kutlaca, A., Turner, D. R., & Morley, A. A. (1992). *In vivo* human somatic mutation: Frequency and spectrum with age. *Mutat. Res.* **266**, 189–96.

Grube, K., & Burkle, A. (1992). Poly (ADP ribose) polymerase activity in mononuclear lymphocytes of 13 mammalian species correlates with species-specific lifespan. *Proc. Natl. Acad. Sci. USA* **89**, 11759–63.

Gupta, R. S. (1980). Senescence of cultured human diploid fibroblasts: Are mutations responsible? *J. Cell. Physiol.* **103**, 209–16.

Gurnell, J. (1987). *The Natural History of Squirrels*. Christopher Helm, London.

Hall, K. Y., Hart, R. W., Benirschke, A. K., & Walford, R. L. (1984). Correlation between ultraviolet-induced DNA repair in primate lymphocytes and fibroblasts and species maximum achievable lifespan. *Mech. Ageing Dev.* **24**, 163–73.

Halliwell, B. (1987). Oxidants and human disease: Some new concepts. *FASEB J.* **1**, 358–64.

Halliwell, B., & Gutteridge, J. M. (1989). *Free Radicals in Biology and Medicine*, 2nd Ed. Oxford University Press, Oxford.

Hamilton, W. D. (1966). The moulding of senescence by natural selection. *J. Theor. Biol.* **12**, 12–45.

Hara, E., Tsurui, H., Shinozaki, A., Nakada, S., & Oda, K. (1991). Cooperative effect of antisense Rb and antisense p53 oligomers on the extension of life span in human diploid fibroblasts, TIG-I. *Biochem. Biophys. Res. Commun.* **179**, 528–34.

Harcourt, A. H., Fossey, D., & Sabater-Pi, J. (1981). Demography of *Gorilla gorilla*. *J. Zool. Lond.* **195**, 215–33.

Hardy, J. (1992). Framing β-amyloid. *Nature Genetics* **1**, 233–4.

Hardy, J., & Mullan, M. (1992). Alzheimer's disease: In search of the soluble. *Nature* **359**, 268–9.

Harley, C. B. (1991). Telomere loss: Mitotic clock or genetic time bomb? *Mutat. Res.* **256**, 271–82.

Harley, C. B., Futcher, A. B., & Greider, C. W. (1990). Telomeres shorten during ageing of human fibroblasts. *Nature* **345**, 458–60.

Harley, C. B., Pollard, J. W., Chamberlain, J. W., Stanners, C. P., & Goldstein, S. (1980). Protein synthetic errors do not increase during the aging of cultured human fibroblasts. *Proc. Natl. Acad. Sci. USA* **77**, 1885–9.

Harman, D. (1956). Aging: A theory based on free radical and radiation chemistry. *J. Gerontol.* **11**, 298–300.

(1981). The aging process. *Proc. Natl. Acad. Sci. USA* **78**, 7124–8.

(1992). Free radical theory of aging. *Mutat. Res.* **275**, 257–66.

Harrison, B. J., & Holliday, R. (1967). Senescence and the fidelity of protein synthesis in *Drosophila*. *Nature* **213**, 990–2.

Harrison, D. E. (1979). Mouse erythropoietic stem cell lines function normally 100 months: Loss related to number of transplantations. *Mech. Ageing Dev.* **9**, 427–33.

(1984). Do hematopoietic stem cells age? *Monogr. Dev. Biol.* **17**, 21–41. Karger, Basel.

(1985). Cell and tissue transplantation: A means of studying the aging process. In *Handbook of the Biology of Aging*, 2nd Ed., ed. C. E. Finch & E. L. Schneider, pp. 322–56. Van Nostrand Reinhold, New York.

Harrison, D. E., Astle, C. M., & Stone, M. (1989). Numbers and functions of transplantable primitive immunohematopoietic stem cells: Effects of age. *J. Immunol.* **142**, 3833–40.

Hart, R. W., & Daniel, F. B. (1980). Genetic stability *in vitro* and *in vivo*. *Adv. Pathobiol.* **7**, 123–41.

Hart, R. W., Sacher, G. A., & Hoskins, T. L. (1979). DNA repair in short- and long-lived rodent species. *J. Gerontol.* **34**, 808–17.

Hart, R. W., & Setlow, R. B. (1974). Correlation between deoxyribonucleic acid excision repair and lifespan in a number of mammalian species. *Proc. Natl. Acad. Sci. USA* **71**, 2169–73.

Hart, R. W., & Turturro, A., Pegram, R. A., & Chou, M. W. (1990). Effects of calorie restriction on the maintenance of genetic fidelity. In *DNA Damage and Repair in Human Tissues*, ed. B. M. Sutherland & A. D. Woodhead, pp. 351–61. Plenum Press, New York.

Hassold, T. [J.], & Chin, D. (1985). Maternal age-specific rates of numerical chromosomal abnormalities with specific reference to trisomy. *Hum. Genet.* **70**, 11–17.

Hassold, T. J., & Jacobs, P. A. (1984). Trisomy in man. *Annu. Rev. Genet.* **18**, 69–97.

Hayflick, L. (1965). The limited *in vitro* lifetime of human diploid cell strains. *Exp. Cell Res.* **37**, 614–36.

(1977). The cellular basis for biological ageing. In *Handbook of the Biology of Ageing*, 1st Ed., ed. C. E. Finch & L. Hayflick, pp. 159–86. Van Nostrand Reinhold, New York.

(1980). Cell aging. *Annu. Rev. Geront. Geriat.* **1**, 26–67.

(1987). The origins of longevity. In *Aging: Modern Biological Theories of Aging*, ed. H. R. Warner, R. N. Butler, R. L. Spratt & E. L. Schnieder, vol. 31, pp. 21–34. Raven Press, New York.

Hayflick, L., & Moorhead, P. S. (1961). The serial cultivation of human diploid cell strains. *Exp. Cell Res.* **25**, 585–621.

Hayssen, V., van Tienhoven, A., & van Tienhoven, A. (1993). *Asdell's Patterns of Mammalian Reproduction*. Cornstock, Cornell University Press, Ithaca.

Hershko, A., & Ciechanover, A. (1992). The ubiquitin system for protein degradation. *Annu. Rev. Biochem.* **61**, 761–807.

Hipkiss, A. R. (1989). The production and removal of abnormal proteins: A key question in the biology of ageing. In *Human Ageing and Later Life*, ed. A. M. Warner, pp. 15–28. Edward Arnold, London.

Hoehn, H., Bryant, E. M., Johnston, P., Norwood, T. H., & Martin, G. M. (1975). Non-selective isolation, stability and longevity of hybrids between normal human somatic cells. *Nature* **258**, 608–10.

Hoehn, H., Bryant, E. M., & Martin, G. M. (1978). The replicative lifespans of euploid hybrids from short-lived and long-lived human skin fibroblast cultures. *Cytogenet. Cell Genet.* **21**, 282–95.

Hoffman, G. W. (1974). On the origin of the genetic code and the stability of the translation process. *J. Mol. Biol.* **86**, 349–62.

Hogan, M. J. (1972). Role of the retinal pigment epithelimum on macular disease. *Ophthalmology* **76**, 6480.

Holehan, A. M., & Merry, B. J. (1986). The experimental manipulation of ageing by diet. *Biol. Rev.* **61**, 329–68.

Holland, J. J., Kohne, D., & Doyle, M. V. (1973). Analysis of virus replication in ageing human fibroblast cultures. *Nature* **245**, 316–19.

Holliday, R. (1964). A mechanism for gene conversion in fungi. *Genet. Res.* **5**, 282–304.

(1969). Errors in protein synthesis and clonal senescence in fungi. *Nature* **221**, 1224–8.

(1975). The growth and death of diploid and transformed human fibroblasts. *Fed. Proc.* **34**, 51–5.

(1979). A new theory of carcinogenesis. *Br. J. Cancer* **40**, 513–22.

(1984a). The significance of DNA methylation in cellular ageing. In *Molecular Biology of Ageing*, ed. A. D. Woodhead, A. D. Blackett, & A. Hollaender, pp. 269–83. Plenum Press, New York.

(1984b). The biological significance of meiosis. In *Controlling Events in Meiosis*, ed. C. W. Evans & H. G. Dickinson, pp. 381–94. Company of Biologists, Cambridge.

(1984c). The unsolved problems of cellular ageing. *Monogr. Dev. Biol.* **17**, 60–77. Karger, Basel.

(1984d). The ageing process is a key problem in biomedical research. *Lancet* **2**, 1386–7.

(1986a). Strong effects of 5-azacytidine on the *in vitro* lifespan of human diploid fibroblasts. *Exp. Cell Res.* **166**, 543–52.

(1987). The inheritance of epigenetic defects. *Science* **238**, 163–70.

(1988a). Towards a biological understanding of the ageing process. *Perspect. Biol. Med.* **32**, 109–23.

(1988b). A possible role for meiotic recombination in germ line reprogramming and maintenance. In *The Evolution of Sex*, ed. R. E. Michot & B. R. Levin, pp. 45–55. Sinauer Associates, Sunderland, Mass.

(1989a). X chromosome reactivation and ageing. *Nature* **337**, 311.

(1989b). Food, reproduction and longevity: Is the extended lifespan of calorie restricted animals an evolutionary adaptation? *BioEssays* **10**, 125–7.

(1990). The limited proliferation of cultured human diploid cells: Regulation or senescence? *J. Gerontol. Biol. Sci.* **45**, B36–41.

(1991a). Quantitative genetic variation and developmental clocks. *J. Theoret. Biol.* **151**, 351–8.

(1991b). A re-examination of the effects of ionising radiation on lifespan and transformation of human diploid fibroblasts. *Mutat. Res.* **256**, 295–302.

(1992). The ancient origins and causes of ageing. *News Physiol. Sci.* **7**, 38–40.

(1993). Epigenetic inheritance based on DNA methylation. In *DNA Methylation: Molecular Biology and Biological Significance*, ed. J. P. Jost & H. P. Saluz, pp. 452–68. Birkhauser, Basel.

(1994). Longevity and fecundity in eutherian mammals. In *Genetics and Evolution of Aging*, ed. M. R. Rose & C. E. Finch. Kluwar Academic Publishers, The Netherlands.

(ed.) (1986b). *Genes, Proteins and Cellular Ageing*. Van Nostrand Reinhold, New York.

Holliday, R., & Grigg, G. W. (1993). DNA methylation and mutation. *Mutat. Res.* **285**, 61–7.

Holliday, R., Huschtscha, L. I., & Kirkwood, T. B. L. (1981). Further evidence for the commitment theory of cellular ageing. *Science* **213**, 1505–8.

Holliday, R., Huschtscha, L. I., Tarrant, G. L., & Kirkwood, T. B. L. (1977). Testing the commitment theory of cellular ageing. *Science* **198**, 366–72.

Holliday, R., & Kirkwood, T. B. L. (1981). Predictions of the somatic mutation and mortalization theories of cellular ageing are contrary to experimental observations. *J. Theor. Biol.* **93**, 627–42.

Holliday, R., Monk, M., & Pugh, J. E. (eds.) (1990). *DNA Methylation and Gene Regulation.* Royal Society, London.

Holliday, R., Porterfield, J. S., & Gibbs, D. D. (1974). Premature ageing and occurrence of altered enzyme in Werner's syndrome fibroblasts. *Nature* **248**, 762–3.

Holliday, R., & Pugh, J. E. (1975). DNA modification mechanisms and gene activity during development. *Science* **187**, 226–32.

Holliday, R., & Rattan, S. I. S. (1984). Evidence that paromomycin induces premature ageing in human fibroblasts. *Monogr. Dev. Biol.* **17**, 221–33. Karger, Basel.

Holliday, R., & Stevens, A. (1978). The effect of an amino acid analogue p-fluorophenylalanine on the longevity of mice. *Gerontology* **24**, 417–25.

Holliday, R., & Tarrant, G. M. (1972). Altered enzymes in ageing human fibroblasts. *Nature* **238**, 26–30.

Holliday, R., & Thompson, K. V. A. (1983). Genetic effects on the longevity of cultured human fibroblasts. III. Correlations with altered glucose-6-phosphate dehydrogenase. *Gerontology* **29**, 89–96.

Hopfield, J. J. (1974). Kinetic proofreading: A new mechanism for reducing errors in biosynthetic processes requiring high specificity. *Proc. Natl. Acad. Sci. USA* **71**, 4135–9.

Hornsby, P. J., & Gill, G. N. (1978). Characterisation of adult bovine adrenocortical cells throughout their life span in tissue culture. *Endocrinology* **102**, 926–36.

Hornsby, P. J., Simonian, M. H., & Gill, G. N. (1979). Aging of adrenocortical cells in culture. *Int. Rev. Cytol.* **10**, 131–62.

Hurwitz, A., & Adashi, E. Y. (1992). Ovarian follicular atresia as an apoptotic process: A paradigm for programmed cell death in endocrine tissues. *Mol. Cell. Endocrinol.* **84**, C19–C23.

Huschtscha, L. I., & Holliday, R. (1983). The limited and unlimited growth of SV40 transformed cells from human diploid MRC-5 fibroblasts. *J. Cell Sci.* **63**, 77–99.

Jacobs, P. A., Brunton, M., & Court Brown, W. M. (1964). Cytogenetic studies in leucocytes on the general population: Subjects of ages 65 years and more. *Ann. Hum. Genet. Lond.* **27**, 353–62.

Jacobs, P. A., Brunton, M., Court Brown, W. M., Doll, R., & Goldstein, H. (1963). Change of human chromosome count distributions with age: Evidence for a sex difference. *Nature* **197**, 1080–1.

Jakubowski, H., & Goldman, E. (1992). Editing of errors in selection of amino acids for protein synthesis. *Microbiol. Rev.* **56**, 412–29.

Joenji, H. (ed.) (1992). Molecular basis of aging: Mitochondrial degeneration and oxidative damage. *Mutat. Res.* **275**, 113–414.

Johnson, R. T. (1986). Anatomy of the aging nerve cell. In *Handbook of the Cell Biology of Aging*, ed. V. J. Cristofalo, pp 149–78. CRC Press, Boca Raton, Fla.

Johnson, T. E. (1987). Aging can be genetically dissected into component processes using long-lived lines of *Caenorhabditis elegans*. *Proc. Natl. Acad. Sci. USA* **84**, 3777–81.

   (1990a). Increased lifespan of *age-1* mutants in *Caenorhabditis elegans* and lower Gompertz rate of aging. *Science* **249**, 908–12.

   (1990b). *Caenorhabditis elegans* offers the potential for molecular dissection of the aging process. In *Handbook of the Biology of Aging*, ed. E. L. Schneider & J. W. Rowe, pp. 45–9. Academic Press, New York.

Jones, M. L. (1982). Longevity of captive mammals. *Zool. Gart. N. F. Jena* **52**, 113–28.

Jongkind, J. F., Verkerk, A., Visser, W. J., & Van Dongen, J. M. (1982). Isolation of autofluorescent 'aged' human fibroblasts by flow sorting. *Exp. Cell Res.* **138**, 409–417.

Jost, J. P., & Saluz, H. P. (eds.) (1993). *DNA Methylation: Molecular Biology and Biological Significance*. Birkhauser, Basel.

Kano, Y., & Little, J. B. (1985). Mechanisms of human cell neoplastic transformation: Relationship of specific abnormal clone formation to prolonged lifespan in X irradiated human diploid fibroblasts. *Int. J. Cancer* **36**, 407–13.

Kanungo, M. S. (1975). A model for ageing. *J. Theor. Biol.* **53**, 253–61.

Kato, H., Harada, M., Tsuchiya, K., & Moriwaki, K. (1980). Absence of correlation between DNA repair in ultraviolet irradiated mammalian cells and lifespan of the donor species. *Jpn. J. Genet.* **55**, 99–108.

Kay, P. H., Pereira, E., Marlow, S. A., Turbett, G., Mitchell, C. A., Jacobson, P. F., Holliday, R., & Papadimitriou, J. M. (1994). Evidence for adenine methylation within the mouse myogenic gene Myo-D1. *Gene* (in press).

Kirkwood, T. B. L. (1977). Evolution of ageing. *Nature* **270**, 301–4.

   (1980). Error propagation in intracellular information transfer. *J. Theor. Biol.* **82**, 363–82.

   (1981). Repair and its evolution: Survival versus reproduction. In *Physiological Ecology: An Evolutionary Approach to Resource Use*, ed. C. R. Townsend & P. Calow, pp. 165–89. Blackwell Scientific, Oxford.

Kirkwood, T. B. L., & Cremer, T. (1982). Cytogerontology since 1881: A reappraisal of August Weismann and a review of modern progress. *Hum. Genet.* **60**, 101–21.

Kirkwood, T. B. L., & Holliday, R. (1975a). The stability of the translation process. *J. Mol. Biol.* **97**, 257–265.

   (1975b). Commitment to senescence: A model for the finite and infinite growth of diploid and transformed human fibroblasts in culture. *J. Theor. Biol.* **53**, 481–96.

   (1979). The evolution of ageing and longevity. *Proc. Roy. Soc. B* **205**, 532–46.

   (1986a). Selection for optimal accuracy and the evolution of ageing. In *Accuracy in Molecular Processes*, ed. T. B. L. Kirkwood, R. F. Rosenberger & D. Gales, pp. 363–79. Chapman & Hall, London.

   (1986b). Ageing as a consequence of natural selection. In *The Biology of Human Ageing*, ed. K. J. Collins & A. H. Bittles, pp. 1–16. Cambridge University Press, Cambridge.

Kirkwood, T. B. L., Holliday, R., & Rosenberger, R. F. (1984). Stability of the cellular translation process. *Int. Rev. Cytol.* **92**, 93–132.

Kirkwood, T. B. L., & Rose, M. R. (1991). Evolution of senescence: Late survival sacrificed for reproduction. *Phil. Trans. Roy. Soc. Lond. B* **332**, 15–24.

Kirkwood, T. B. L., Rosenberger, R. F., & Galas, D. J. (1986) (eds.) *Accuracy in Molecular Processes*. Chapman & Hall, London.

Kornberg, A., & Baker, T. A. (1992). *DNA Replication*, 2nd Ed. W. H. Freeman, San Francisco.

Kowald, A., & Kirkwood, T. B. L. (1993). Accuracy of tRNA charging and codon : anticodon recognition; relative importance for cellular stability. *J. Theor. Biol.* **160**, 493–508.

Krohn, P. L. (1962). Heterochronic transplantation in the study of ageing. *Proc. Roy. Soc. B* **157**, 128–47.

Kurland, C. G. (1987). The error catastrophe: A molecular *fata morgana*. *Bio-Essays* **6**, 33–5.

Land, H., Parada, L. F., & Weinberg, R. A. (1983). Tumorigenic conversion of primary embryo fibroblasts requires at least two cooperating oncogenes. *Nature* **304**, 596–602.

Lange, C. S. (1968). Studies on the cellular basis of radiation lethality. I. The pattern of mortality in the whole-body irradiated planarian (*Tricladida, Paludicola*). *Int. J. Radiat. Biol.* **13**, 511–30.

Lansing, A. I. (1947). A transmissable, cumulative and reversible factor in ageing. *J. Gerontol.* **2**, 228–39.

Lechner, J. F., Haugen, A., Autrup, H., McClendon, I. A., Trump, B. F., & Harris, C. C. (1981). Clonal growth of epithelial cells from normal adult human bronchus. *Cancer Res.* **41**, 2294–304.

Lee, A. T., & Cerami, A. (1990). Modifications of proteins and nucleic acids by reducing sugars: possible role in aging. In *Handbook of the Biology of Aging*, 3rd Ed., ed. E. L. Schneider & J. W. Rowe, pp. 116–30. Academic Press, San Diego.

Levine, A. J. (1993). The tumour suppressor genes. *Annu. Rev. Biochem.* **62**, 623–51.

Levy, M. Z., Allsopp, R. C., Futcher, A. B., Greider, C. W., & Harley, C. B. (1992). Telomere end-replication problem and cell ageing. *J. Mol. Biol.* **225**, 951–60.

Lewis, C. M., & Holliday, R. (1970). Mistranslation and ageing in *Neurospora*. *Nature* **288**, 877–80.

Libby, R. T., & Gallant, J. A. (1991). The role of RNA polymerase in transcriptional fidelity. *Mol. Microbiol.* **5**, 999–1004.

Libby, R. T., Nelson, J. L., Calvo, J. M., & Gallant, J. A. (1989). Transcriptional proofreading in *Escherichia coli. EMBO J.* **8**, 3153–8.

Lindahl, T. (1979). DNA glycosylases, endonucleases for apurinic/pyrimidinic sites and base excision-repair. *Prog. Nucleic Acid Res. Mol. Biol.* **22**, 135–92.

(1993). Instability and decay of the primary structure of DNA. *Nature* **362**, 709–15.

Lindop, P. J., & Rotblat, J. (1961). Shortening of life and causes of death in mice exposed to single whole-body dose of radiation. *Nature* **189**, 645–8.

Lindsay, J., McGill, N. I., Lindsay, L. A., Green, D. K., & Cooke, H. J. (1991). *In vivo* loss of telomeric repeats with age in humans. *Mutat. Res.* **256**, 45–8.

Linn, S., Kairis, M., & Holliday, R. (1976). Decreased fidelity of DNA polymerase activity isolated from ageing human fibroblasts. *Proc. Natl. Acad. Sci. USA* **73**, 2818–22.

Linnane, A. W., Marzuki, S., Ozawa, T., & Tanaka, M. (1989). Mitochondrial DNA mutations as an important contribution to ageing and degenerative diseases. *Lancet* **1**, 642–5.

Lipetz, J., & Cristofalo, V. J. (1972). Ultrastructural changes accompanying the aging of human diploid cells in culture. *J. Ultrastruct. Res.* **39**, 43–56.

Loftfield, R. B. (1963). The frequency of errors in protein synthesis. *Biochem. J.* **89**, 82–92.

Loftfield, R. B., & Vanderjagt, D. (1972). The frequency of errors in protein biosynthesis. *Biochem. J.* **128**, 1353–6.

Lu, A. Y. H., & West, S. B. (1979). Multiplicity of mammalian microsomal cytochromes P-450. *Pharmacol. Rev.* **31**, 277–95.

Luce, M. C., & Bunn, C. L. (1989). Decreased accuracy of protein synthesis in extracts from ageing human diploid fibroblasts. *Exp. Gerontol.* **24**, 113–25.

Lundblad, V., & Szostak, J. W. (1989). A mutant with a defect in telomere elongation leads to senescence in yeast. *Cell* **57**, 633–43.

McCarron, M., Osborne, Y., Story, C. J., Dempsey, J. L., Turner, D. R., & Morley, A. A. (1987). Effect of age or lymphocyte proliferation. *Mech. Ageing Dev.* **41**, 211–18.

McCay, C. M., Maynard, L. A., Sperling, G., & Barnes, L. L. (1939). Retarded growth, lifespan, ultimate body size and age changes in the albino rat after feeding diets restricted in colonies. *J. Nutr.* **18**, 1–13.

McFarland, G. M., & Holliday, R. (1994). Retardation of the senescence of cultured human diploid fibroblasts by carnosine. *Exp. Cell Res.* **212** (in press).

Macieira-Coelho, A. (1966). Action of cortisone on human fibroblasts *in vitro*. *Experientia* **22**, 390–1.

Macieira-Coelho, A., Dietloff, C., Billaidon, C., Bourgeous, C. A., & Malaise, E. (1977). Effect of low dose rate ionising radiation on the division potential of cells *in vitro*. III. Human lung fibroblasts. *Exp. Cell Res.* **104**, 215–21.

McKusick, V. A. (1990). *Mendelian Inheritance in Man*, 9th Ed. Johns Hopkins University Press, Baltimore.

McLaren, A. (1992). The quest for immortality. *Nature* **359**, 482–3.

Makinodan, T. (1979). Control of immunologic abnormalities associated with aging. *Mech. Ageing Dev.* **9**, 7–17.

Makinodan, T., Allbright, J. W., Good, P. I., Peter, C. P., & Heidrick, M. L. (1976). Reduced humoral immune activity in long-lived old mice: An approach to elucidating its mechanisms. *Immunology* **31**, 903–11.

Makinodan, T., & Kay, M. M. B. (1980). Age influence on the immune system. *Adv. Immunol.* **29**, 287–330.

Makrides, S. (1983). Protein synthesis and degradation during ageing and senescence. *Biol. Rev.* **58**, 343–422.

Marcou, D. (1961). Notion de longévité et nature cytoplasmique due déterminant de la sénscence chez quelques champignons. *Ann. Sci. Nat. Bot. Biol. Veg.* **12**, 653–764.

Martin, G. M. (1978). Genetic syndromes in man with potential relevance to the pathology of ageing. In *Genetic Effects on Ageing*, ed. D. Bergsma & D. E. Harrison. Alan Liss, New York.

Martin, G. M., Curtis, A., Sprague, B. S., & Epstein, C. J. (1970). Replicative lifespan of cultivated human cells: Effects of donor's age, tissue and genotype. *Lab. Invest.* **23**, 86–92.

Masoro, E. J. (ed.) (1981). *Handbook of Physiology in Aging*. CRC Press, Boca Raton, Fla.

Matsumura, T., Hunter, J., Malik, F., & Holliday, R. (1989). Maintenance of

DNA methylation level in SV40 infected human fibroblasts during their *in vitro* limited proliferative lifespan. *Exp. Cell Res.* **184,** 148–57.

Maynard Smith, J. (1959). A theory of ageing. *Nature* **184,** 956–7.

(1962). The causes of ageing. *Proc. Roy. Soc. B* **157,** 115–27.

Mays Hoopes, L. L. (1989). Age related changes in DNA methylation: Do they represent continued developmental changes? *Int. Rev. Cytol.* **114,** 181–220.

Medawar, P. B. (1952). *An Unsolved Problem in Biology.* Lewis, London. Reprinted in Medawar, P. B. (1981) *The Uniqueness of the Individual.* Dover, New York.

Medvedev, Zh. A. (1962) Ageing at the molecular level and some speculations concerning maintaining the functioning of systems for replicating specific macromolecules. In *Biological Aspects of Ageing,* ed. N. Shock, pp. 255–66. Columbia University Press, New York.

(1980). The role of infidelity of transfer of information for the accumulation of age changes in differentiated cells. *Mech. Ageing Dev.* **14,** 1–14.

(1990). An attempt at a rational classification of theories of aging. *Biol. Rev.* **65,** 375–98.

Medvedev, Zh. A., & Medvedeva, M. (1991). Age changes in chromatin: Accumulation or programmed? *Ann. New York Acad. Sci.* **621,** 40–52.

Menninger, J. R. (1977). Ribosome editing and the error catastrohe hypothesis of cellular ageing. *Mech. Ageing Dev.* **10,** 379–98.

Metchnikoff, E. (1907). *The Prolongation of Life.* Heinemann, London.

Michaelis, J., & Hipkiss, A. R. (1994). Non-enzymic glycosylation of the dipeptide carnosine. *Biochim. Biophys. Acta,* submitted for publication.

Migeon, B. R., Axelman, J., & Beggs, A. H. (1988). Effect of ageing on reactivation of the human X-linked HPRT locus. *Nature* **335,** 93–6.

Milisauskas, V., & Rose, N. R. (1973). Immunochemical quantitation of enzymes in human diploid cell line W1-38. *Exp. Cell Res.* **81,** 279–84.

Miller, R. A. (1990). Aging and the immune response. In *Handbook of the Biology of Aging,* ed. E. L. Schneider & J. W. Rowe, pp. 157–80. Academic Press, New York.

Modrich, P. (1991). Mechanisms and biological effects of mismatch repair. *Annu. Rev. Genet.* **25,** 229–53.

Monk, M., Boubelik, M., & Lehnert, S. (1987). Temporal and regional changes in DNA methylation in the embryonic, extraembryonic and germ cell lineages during mouse embryo development. *Development* **99,** 371–82.

Monk, M., & Surani, A. (eds.) (1990). *Genomic Imprinting.* Development Suppl. 1990. Company of Biologists, Cambridge.

Monnat, R. J. (1992). Werner syndrome: Molecular genetics and mechanistic hypotheses. *Exp. Gerontol.* **27,** 447–53.

Monnier, V. M. (1988). Towards a Maillard reaction theory of aging. In *The Maillard Reaction in Aging,* ed. S. W. Baynes & V. M. Monnier, pp. 1–22. Alan Liss, New York.

Monnier, V. M., Kohn, R. R., & Cerami, A. (1984). Accelerated age-related browning of human collagen in diabetes mellitus. *Proc. Natl. Acad. Sci. USA* **81,** 583–7.

Monti, D., Troiano, L., Tropea, F., Grassilli, E., Cossarizza, A., Barozzi, D., Pelloni, M-C., Tamassia, M. G., Bellomo, G., & Franceschi, C. (1992). Apoptosis-programmed cell death: A role in the aging process? *Am. J. Clin. Nutr.* **55,** 1208S–14S.

Morley, A. A. (1982). Is ageing the result of dominant and co-dominant mutations? *J. Theor. Biol.* **98,** 469–74.
Morley, A. A., Cox, S., & Holliday, R. (1982). Human lymphocytes resistant to
6-thioguanine increases with age. *Mech. Ageing Dev.* **19,** 21–6.
Morley, A. A., Trainor, K. J., Seshadri, R., & Ryall, R. G. (1983). Measurements
of *in vivo* mutations in human lymphocytes. *Nature* **302,** 155–6.
Morris, J. H. (1989). The nervous system. In *Robbins' Pathologic Basis of Disease*, 4th Ed., pp. 1385–1449. W. B. Saunders, Philadelphia.
Mueller, S. N., Rosen, E. M., & Levine, E. M. (1980). Cellular senescence in a
cloned strain of bovine fetal aortic endothelial cells. *Science* **207,** 889–
91.
Murray, V. (1990) Are transposons a cause of ageing? *Mutat. Res.* **237,** 59–63.
Murray, V., & Holliday, R. (1979a). A mechanism for RNA–RNA splicing and
a model for the control of gene expression. *Genet. Res.* **34,** 173–88.
    (1979b). Mechanism for RNA splicing of gene transcripts. *FEBS Lett.* **106,**
    5–7.
    (1981). Increased error frequency of DNA polymerases from senescent hu-
    man fibroblasts. *J. Mol. Biol.* **146,** 55–76.
Nagel, J. E., Yanagihara, R. H., & Adler, W. H. (1986). Cells of the immune re-
sponse. In *Handbook of the Cell Biology of Aging*, ed. V. J. Cristofalo, pp.
341–63. CRC Press, Boca Raton, Fla.
Neary, G. J. (1960). Ageing and radiation. *Nature* **187,** 10–18.
Nebert, D. W., & Gonzalez, F. J. (1987). P450 genes: structure evolution and
regulation. *Annu. Rev. Biochem.* **56,** 945–93.
Ninio, J. (1975). Kinetic amplification of enzyme discrimination. *Biochimie* **57,**
587–95.
Norwood, T. H., Pendergrass, W. R., Sprague, C. A., & Martin, G. M. (1974).
Dominance of the senescent phenotype in heterokaryons between replica-
tive and post-replicative human fibroblast cells. *Proc. Natl. Acad. Sci. USA*
71, 2231–4.
Norwood, T. H., Smith, J. R., & Stein, G. H. (1990). Aging at the cellular level:
The human fibroblast-like cell model. In *Handbook of the Biology of Ag-
ing*, 3rd Ed., ed. E. L. Schneider & J. W. Rowe, pp. 131–54. Academic
Press, New York.
Nowak, R. M. (1991). *Walker's Mammals of the World,* 5th Ed. Johns Hopkins
Press, Baltimore.
Nunn, P. (1994). Chemical toxins. In *Motor Neuron Disease*, ed. C. A. Williams,
pp. 567–6. Chapman & Hall, London.
Ogden, D. A., & Micklem, H. S. (1976). The fate of serially transplanted bone
marrow cell populations from young and old donors. *Transplantation* **22,**
287–93.
Oliver, C. N., Ahn, B. W., Moerman, E. J., Goldstein, S., & Stadtman, E. R.
(1987). Age-related changes in oxidised proteins. *J. Biol. Chem.* **262,**
5488–91.
Olovnikov, A. M. (1973). A theory of marginotomy. *J. Theor. Biol.* **41,** 181–
90.
Olson, C. B. (1987). A review of why and how we age: A defence of multifactorial aging. *Mech. Ageing Dev.* **41,** 1–28.
Orgel, L. E. (1963). The maintenance of the accuracy of protein synthesis and
its relevance to ageing. *Proc. Natl. Acad. Sci. USA* **49,** 517–21.
    (1970). The maintenance of the accuracy of protein synthesis and its rele-
    vance to ageing: A correction. *Proc. Natl. Acad. Sci. USA* **67,** 1476.

(1973). Ageing of clones of mammalian cells. *Nature* **243**, 441–5.

Parke, D. V., Ioannides, C., & Lewis, D. F. V. (1991). The role of cytochromes P450 in the detoxification and activation of drugs and other chemicals. *Can. J. Physiol. Pharmacol.* **69**, 537–49.

Parker, J., Pollard, J. W., Friesen, J. D., & Stanners, C. P. (1978). Stuttering: High level mistranslation in animal and bacterial cells. *Proc. Natl. Acad. Sci. USA* **75**, 1091–5.

Partridge, L., & Barton, N. H. (1993). Optimality, mutation and the evolution of ageing. *Nature* **362**, 305–11.

Pauling, L. (1958). *Festchr. Arthur Stoll*, p. 597, Birkhauser, Basel.

Peleg, L., Raz, E., & Ben-Ishai, R. (1977). Changing capacity for DNA excision repair in mouse embryonic cells *in vitro*. *Exp. Cell Res.* **104**, 301–7.

Pereira-Smith, O. M., & Smith, J. R. (1982). Phenotype of low proliferative potential is dominant in hybrids of normal human fibroblasts. *Somatic Cell Genet.* **8**, 731–42.

(1983). Evidence for the recessive nature of cellular immortality. *Science* **221**, 964–6.

Pero, R. W., Holmgren, K., & Persson, L. (1985). γ-radiation induced ADP-ribosyl transferase activity and mammalian longevity. *Mutat. Res.* **142**, 69–73.

Petes, T. D., Farber, R. A., Tarrant, G. M., & Holliday, R. (1974). Altered rate of DNA replication in ageing human fibroblast cultures. *Nature* **251**, 434–6.

Peto, R. (1977) Epidemiology, multistage models and short-term mutagenicity tests. In *Origins of Human Cancer*, ed. H. H. Hiatt, J. D. Watson & J. A. Winsten, pp. 1403–28. Cold Spring Harbor Laboratory Press, New York.

Peto, R., Parish, S. E., & Gray, R. G. (1986). There is no such thing as ageing, and cancer is not related to it. *IARC Publications* **58**, 43–53.

Peto, R., Roe, F. J. C., Lee, P. N., Levy, L., & Clack, J. (1975). Cancer and ageing in mice and men. *Brit. J. Cancer* **32**, 411–26.

Ponten, J. (1973). Aging properties of human glia. *INSERM* **27**, 53–64.

Ponten, J., & MacIntyre, E. H. (1968). Long term culture of normal and neoplastic human glia. *Acta Pathol. Microbiol. Scand.* **74**, 465–86.

Potten, C. S., & Loeffler, M. (1990). Stem cells: Attributes, cycles, spirals, pitfalls and uncertainties. Lessons for and from the crypt. *Development* **110**, 1001–20.

Printz, D. B., & Gross, S. R. (1967). An apparent relationship between mistranslation and an altered leucyl tRNA synthetase in a conditional lethal mutant of *Neurospora crassa*. *Genetics* **55**, 451–67.

Prusiner, S. B., Collinge, J., Powell, B., & Anderton, B. (eds.) (1993). *Prion Diseases of Humans and Animals*. Ellis Horwood, London.

Randerath, E., Hart, R. W., Turturro, A., Danna, T. F., Reddy, R., & Randerath, K. (1991). Effects of aging and calorie restriction on I compounds in liver, kidney and white blood cell DNA of male Brown-Norway rats. *Mech. Ageing Dev.* **58**, 279–96.

Rattan, S. I. S. (1989). DNA damage and repair during cellular ageing. *Int. Rev. Cytol.* **116**, 47–88.

Rattan, S. I. S., Derventzi, A., & Clark, B. F. C. (1992). Protein synthesis, post-translation modifications, and aging. *Ann. New York Acad. Sci.* **663**, 48–62.

Rattan, S. I. S., Keeler, K. D., Buchanan, J. H., & Holliday, R. (1982). Autofluorescence as an index of ageing in human fibroblasts in culture. *Biosci. Rep.* **2**, 561–7.

Razin, A., & Cedar, H. (1993). DNA methylation and embryogenesis. In *DNA Methylation: Molecular Biology and Biological Significance*, ed. J. P. Jost & H. P. Saluz, pp. 343–57. Birkhauser, Basel.

Rechsteiner, M. (1987). Ubiquitin-mediated pathways for intracellular proteolysis. *Annu. Rev. Cell Biol.* **3**, 1–30.

(1991). Natural substrates of the ubiquitin proteolytic pathway. *Cell* **66**, 615–18.

Richter, C., Park, J. W., & Ames, B. N. (1988). Normal oxidative damage to mitochondrial and nuclear DNA is extensive. *Proc. Natl. Acad. Sci. USA* **85**, 6465–7.

Riggs, A. D. (1975). X inactivation, differentiation and DNA methylation. *Cytogenet. Cell Genet.* **14**, 9–25.

Risch, N., Reich, E. W., Wishnick, M. M., & McCarthy, J. G. (1987). Spontaneous mutation and parental age in humans. *Am. J. Hum. Genet.* **41**, 218–48.

Robbins, E., Levine, E. M., & Eagle, H. (1970). Morphologic changes accompanying senescence of cultured human diploid cells. *J. Exp. Med.* **131**, 1211–22.

Robinson, A. B., McKerron, J. H., & Cary, P. (1970). Controlled deamidation of peptides and proteins: An experimental hazard and a possible biological timer. *Proc. Natl. Acad. Sci. USA* **66**, 753–7.

Rockwell, G. A., Johnson, G., & Sibatani, A. (1987). *In vitro* senescence of human keratinocyte cultures. *Cell Struct. Funct.* **12**, 539–48.

Rohme, D. (1981). Evidence for a relationship between longevity of mammalian species and lifespans of normal fibroblasts *in vitro* and erythrocytes *in vivo*. *Proc. Natl. Acad. Sci. USA* **78**, 5009–13.

Roitt, I. (1988). *Essential Immunology*, 6th Ed. Blackwell, Oxford.

Rose, M. R. (1984). Laboratory evolution of postponed senescence in *Drosophila melanogaster*. *Evolution* **38**, 1004–10.

(1991). *Evolutionary Biology of Aging*. Oxford University Press, Oxford.

Rosenberger, R. F. (1982). Streptomycin-induced protein error propagation appears to lead to cell death in *Escherichia coli*. *IRCS Med. Sci.* **10**, 874–5.

(1991). Senescence and the accumulation of abnormal proteins. *Mutat. Res.* **256**, 243–54.

Rosenberger, R. F., Gounaris, E., & Kolettas, E. (1991). Mechanisms responsible for the limited lifespan and immortal phenotypes in cultured mammalian cells. *J. Theor. Biol.* **148**, 383–92.

Rothstein, M. (1975). Ageing and alteration of enzymes: A review. *Mech. Ageing Dev.* **4**, 325–38.

(1977). Recent developments in the age-related alteration of enzymes: A review. *Mech. Ageing Dev.* **6**, 241–57.

(1979). The formation of altered enzymes in ageing animals. *Mech. Ageing Dev.* **9**, 197–202.

(1982). *Biochemical Approaches to Aging*. Academic Press, New York.

Ryan, J. M., Duda, G., & Cristofalo, V. J. (1974). Error accumulation and ageing in human diploid cells. *J. Gerontol.* **29**, 616–21.

Sacher, G. A., & Hart, R. W. (1978). Longevity, aging and comparative cellular and molecular biology of the house mouse, *Mus musculus,* and the white-footed mouse, *Peromyscus leucopus*. In *Genetic Effects on Aging*, ed. D. Bergsma & D. E. Harrison, pp. 71–96. Alan Liss, New York.

Saksela, E., & Moorhead, P. S. (1963). Aneuploidy in the degenerative phase of serial cultivation of human cell strains. *Proc. Natl. Acad. Sci. USA* **50**, 390–5.

Salk, D., Au, K., Hoehn, H., & Martin, G. M. (1981). Cytogenetics of Werner's syndrome cultured skin fibroblasts: Variegated translocation mosaicism. *Cytogenet. Cell Genet.* **30**, 92–107.

Salk, D., Fujiwara, Y., & Martin, G. M. (eds.) (1985). *Werner's Syndrome and Human Ageing.* Plenum, New York.

Saxton, G. A., & Servadda, D. M. (1969). Human birth interval in East Africa. *J. Reprod. Fertil.* **6**, 83–8.

Schneider, E. L., & Reed, J. D. (1985). Modulation of aging processes. In *Handbook of the Biology of Aging*, 2nd Ed., ed. C. E. Finch & E. L. Schneider, pp. 45–76. Van Nostrand Reinhold, New York.

Schrodinger, E. (1944). *What Is Life?* Cambridge University Press, Cambridge.

Schwartz, A. G. (1975) Correlation between species lifespan and capacity to activate 7,12 dimethyl benzanthracene to a form mutagenic to a mammalian cell. *Exp. Cell Res.* **94**, 445–7.

Schwartz, A. G., & Moore, C. J. (1977). Inverse correlation between species lifespan and capacity of cultured fibroblasts to bind 7,12 dimethyl benzanthracene to DNA. *Exp. Cell Res.* **109**, 448–50.

Sedgwick, S. G. (1986). Stability and change through DNA repair. In *Accuracy in Molecular Processes*, ed. T. B. L. Kirkwood, R. F. Rosenberger & D. J. Galas, pp. 233–89. Chapman & Hall, London.

Shakespeare, V. A., & Buchanan, J. H. (1976). Increased degradation rates of protein in ageing human fibroblasts and in cells treated with an amino acid analogue. *Exp. Cell Res.* **100**, 1–8.

(1979). Increased proteolytic activity in ageing human fibroblasts. *Gerontology* **25**, 305–13.

Shay, J. W., Wright, W. E., & Werbin, H. (1991). Defining the molecular mechanisms of human cell immortalisation. *Biochim. Biophys. Acta* **1072**, 1–7.

Sheng, T. C. (1951). A gene that causes natural death in *Neurospora crassa. Genetics* **36**, 199–212.

Sies, H. (1993). Strategies of antioxidant defence. *Eur. J. Biochem.* **215**, 213–19.

Smith, C. W., Patton, J. G., & Nadal-Ginard, B. (1989). Alternative splicing in the control of gene expression. *Annu. Rev. Genet.* **23**, 527–77.

Smith, J. R. (1984). An hypothesis for *in vitro* cellular senescence based on the population dynamics of human diploid fibroblasts and somatic cell hybrids. *Monogr. Dev. Biol.* **17**, 193–207. Karger, Basel.

(1990). DNA synthesis inhibitors in cellular senescence. *J. Gerontol.* **45**, 32–5.

Smith, J. R., & Lincoln, D. W. (1984). Aging of cells in culture. *Int. Rev. Cytol.* **89**, 151–77.

Smith, J. R., & Lumpkin, C. K. (1980). Loss of gene repression activity: A theory of cellular senescence. *Mech. Ageing Dev.* **13**, 387–92.

Smith, J. R., Pereira Smith, O. M., & Schneider, E. L. (1978). Colony size distribution as a measure of *in vivo* and *in vitro* ageing. *Proc. Natl. Acad. Sci. USA* **75**, 1353–6.

Smith, J. R., & Whitney, R. G. (1980). Intraclonal variation in proliferative potential of human diploid fibroblasts: Stochastic mechanism for cellular aging. *Science* **207**, 82–4.

Sonneborn, T. M. (1930). Genetic studies on *Stenostomum incaudatum* (nov. sp.). I. The nature and origin of differences among individuals formed during vegetative reproduction. *J. Exp. Zool.* **57**, 57–108.

Stadtman, E. R. (1992) Protein oxidation and aging. *Science* **257**, 1220–4.

Stein, G. H., & Atkins, L. (1986). Membrane-associated inhibitor of DNA synthesis in senescent human fibroblasts: Characterisation and comparison to quiescent cell inhibitor. *Proc. Natl. Acad. Sci. USA* **83,** 9030–4.

Stenback, F., Peto, R., & Shubik, P. (1981). Initiation and promotion at different ages and doses in 2200 mice. *Br. J. Cancer* **44,** 1–31.

Stieglitz, E. J. (1942). The social urgency of research in ageing. In *Problems of Ageing: Biological and Medical Aspects*, 2nd Ed., ed. E. V. Cowdray, pp. 890–907, Williams & Williams, Baltimore.

Sulston, J. E., Schierenberg, E., White, J. G., & Thomson, J. N. (1983). The embryonic cell lineage of the nematode *Caenorhabditis elegans. Dev. Biol.* **100,** 64–119.

Szilard, L. (1959). On the nature of the ageing process. *Proc. Natl. Acad. Sci. USA* **45,** 35–45.

Takeda, T., Hosokawa, M., & Higuchi, K. (1991). Senescence-accelerated mouse (SAM): A novel murine model of accelerated senescence. *J. Am. Geriatr. Soc.* **39,** 911–19.

Tanzi, R. E., St. George-Hyslop, P., & Gusella, J. F. (1991). Molecular genetics of Alzheimer disease amyloid. *J. Biol. Chem.* **266,** 20579–82.

Thompson, K. V. A., & Holliday, R. (1973). Effect of temperature on the longevity of human fibroblasts in culture. *Exp. Cell Res.* **80,** 354–60.

(1975). Chromosome changes during the *in vitro* ageing of MRC-5 human fibroblasts. *Exp. Cell Res.* **96,** 1–6.

(1978). The longevity of diploid and polyploid human fibroblasts: Evidence against the somatic mutation theory of cellular ageing. *Exp. Cell Res.* **112,** 281–7.

(1983). Genetics effects on the longevity of cultured human fibroblasts. I. Werner's syndrome. *Gerontology* **29,** 73–82.

Tice, R. R., & Setlow, R. B. (1985). DNA repair and replication in aging organisms and cells. In *Handbook of the Biology of Aging*, 2nd Ed., ed. C. E. Finch & E. L. Schneider, pp. 173–224. Van Nostrand Reinhold, New York.

Tilly, J. L., Kowalski, K. I., Johnson, A. L., & Hsueh, A. J. W. (1991). Involvement of apoptosis in ovarian follicular atresia and post-ovulating regression. *Endocrinology* **129,** 2799–801.

Todaro, G. J., & Green, H. (1963). Quantitative studies of the growth of mouse embryo cells in culture and their development into established lines. *J. Cell Biol.* **17,** 299–313.

Tollefsbol, T. O., Zaun, M. R., & Gracy, R. W. (1982). Increased lability of triosephosphate isomerase in progeria and Werner's syndrome fibroblasts. *Mech. Ageing Dev.* **20,** 93–101.

Tolmasoff, J. M., Ono, T., & Cutler, R: G. (1980). Superoxide dismutase: Correlation with lifespan and specific metabolic rate in primate species. *Proc. Natl. Acad. Sci. USA* **77,** 2777–81.

Tomei, L. D., & Cope, F. O. (eds.) (1991). *Apoptosis: The Molecular Basis of Cell Death*. Cold Spring Harbor Laboratory Press, New York.

Trainor, K. J., Wigmore, D. J., Chrysostomou, A., Dempsey, J. L., Seshadri, R., & Morley, A. A. (1984). Mutation frequency of human lymphocytes increases with age. *Mech. Ageing Dev.* **27,** 83–6.

Treton, J. A., & Courtois, Y. (1982). Correlation between DNA excision repair and mammalian lifespan in lens epithelial cells. *Cell Biol. Int. Rep.* **6,** 253–60.

Turturro, A., & Hart, R. W. (1991). Longevity assurance mechanisms and calorie restriction. *Ann. New York Acad. Sci.* **621**, 363–72.

van Boekel, M. A. M. (1991). The role of glycation in aging and diabetes mellitus. *Mol. Biol. Rep.* **15**, 57–64.

Vincent, R. A., & Huang, P. C. (1976). The proportion of cells labeled with tritiated thymidine as a function of population doubling level in cultures of fetal, adult, mutant and tumor origin. *Exp. Cell Res.* **102**, 31–42.

Wagner, J. R., Hu, C.-C., & Ames, B. N. (1992). Endogenous oxidative damage of deoxycytidine in DNA. *Proc. Natl. Acad. Sci. USA* **89**, 3380–4.

Walford, R. L. (1969). *The Immunologic Theory of Ageing.* Munksgaard, Copenhagen.

—— (1974). Immunologic theory of aging: Current status. *Fed. Proc.* **33**, 2020–7.

Wang, K-M., Rose, N. R., Bartholemew, E. A., Balzer, M., Berde, K., & Foldvary, M. (1970). Changes of enzymatic activities in human diploid line W1-38 at various passages. *Exp. Cell Res.* **61**, 357–64.

Wareham, K. A., Lyon, M. F., Glenister, P. H., & Williams, E. D. (1987). Age related reactivation of an X-linked gene. *Nature* **327**, 725–7.

Warner, H. R., Butler, R. N., Spratt, R. L., & Schneider, E. L. (eds.) (1987). *Modern Biological Theories of Aging.* Raven Press, New York.

Warner, H. R., & Wang, E. (eds.) (1989). Proceedings of the Workshop on *Control of Cell Proliferation in Cells. Exp. Gerontol.* **24**, 351–585.

Wei, Q., Matanoski, G. M., Farmer, E. R., Hedayati, M. A., & Grossman, L. (1993). DNA repair and aging in basal cell carcinoma: A molecular epidemiology study. *Proc. Natl. Acad. Sci. USA* **90**, 1614–18.

Weintraub, H. (1985). Assembly and propagation of repressed and derepressed chromosome states. *Cell* **42**, 705–11.

Wesson, L. C. (1969). *Physiology of the Human Kidney.* Grune & Stratton, New York.

Wiebauer, K., & Jiricny, J. (1989). *In vitro* correction of G·T mispairs to G·C pairs in nuclear extracts from human cells. *Nature* **339**, 234–6.

—— (1990). Mismatch-specific thymine DNA glycosylase and DNA polymerase β mediate the correction of GT mispairs in nuclear extracts from human cells. *Proc. Natl. Acad. Sci. USA* **87**, 5842–5.

Williams, D. D., Short, R., & Bowden, M. B. (1990). Fingernail growth rate as a biomarker of aging in the pigtailed macaque (*Macaca nemestrina*). *Exp. Gerontol.* **25**, 423–32.

Williams, G. C. (1957). Pleiotropy, natural selection and the evolution of senescence. *Evolution* **11**, 398–411.

Williamson, A. R., & Askonas, B. A. (1972). Senescence of an antibody-forming cell clone. *Nature* **238**, 337–9.

Wilson, A. C. (1991). From molecular evolution to body and brain evolution. In *Perspectives on Cellular Regulation: From Bacteria to Cancer.* MBL Lectures in Biology, vol. 11, pp. 331–40. John Wiley, New York.

Wilson, V. L., & Jones, P. A. (1983). DNA methylation decreases in ageing but not in immortal cells. *Science* **220**, 1055–7.

Wilson, V. L., Smith, R. A., Ma, S., & Cutler, R. G. (1987). Genomic 5-methyl deoxycytidine decreases with age. *J. Biol. Chem.* **262**, 9948–51.

Witkowski, J. A. (1980). Dr. Carrel's immortal cells. *Med. Hist.* **24**, 129–42.

—— (1987). Cell aging *in vitro*: A historical perspective. *Exp. Gerontol.* **22**, 231–48.

Wodinsky, J. (1977). Hormonal inhibition of feeding and death in *Octopus:* Control by optic gland secretion. *Science* **198,** 948–51.

World Health Organization (1991). *World Health Statistics,* p. 10. WHO, Geneva.

Wyllie, A. H. (1992). Apoptosis and the regulation of cell numbers in normal and neoplastic tissues. *Cancer Metastasis Rev.* **11,** 95–103.

Yamauchi, M., Woodley, D. T., & Mechanic, G. L. (1988). Aging and cross-linking of skin collagen. *Biochem. Biophys. Res. Commun.* **152,** 898–903.

Yuan, P. M., Talent, J. P., & Gracy, R. W. (1981). Molecular basis for the accumulation of acid isozymes of triosephosphate isomerase on ageing. *Mech. Ageing Dev.* **17,** 151–62.

Youngman, L. D., Park, J-Y. K., & Ames, B. N. (1992). Protein oxidation associated with aging is reduced by dietary restriction of proteins or calories. *Proc. Natl. Acad. Sci. USA* **89,** 9112–16.

Zavala, C., Herner, G., & Fialkow, P. F. (1978). Evidence for selection in cultured diploid fibroblast strains. *Exp. Cell Res.* **117,** 137–44.

Zurcher, C., van Zwieten, M. J., Solleveld, H. A., & Hollander, C. F. (1982). Ageing research. In *The Mouse in Biomedical Research*, vol. IV, ed. H. L. Foster, J. D. Small & J. G. Fox, pp. 11–35. Academic Press, New York.

Author index

Subject index

198